INDIANS, MARKETS, AND RAINFORESTS

INDIANS, MARKETS, AND RAINFORESTS

THEORY, METHODS, ANALYSIS

Ricardo A. Godoy

Columbia University Press
New York

Columbia University Press
Publishers Since 1893
New York   Chichester, West Sussex

Chapter 11 appeared in slightly different form as "Of trade and cognition: Markets
and the loss of folk knowledge among the Tawahka indians of the Honduran rain forest."
Journal of Anthropological Research 54(1998):219–33. Used by permission.

Library of Congress Cataloging-in-Publication Data

Godoy, Ricardo A., 1951-
Indians, markets, and rainforests : theory, methods, analysis / Ricardo Godoy.
p.   cm.
Includes bibliographical references and index.
ISBN 978-0-231-11784-5 (cloth : alk. paper) — 978-0-231-11785-2 (pbk. : alk. paper)
1. Indians of Central America—Social conditions. 2. Indians of South America—
Social conditions. 3. Rain forests—Economic aspects—Latin America. 4. Rain forest
ecology—Latin America. 5. Human ecology—Government policy—Latin America.
6. Deforestation—Economic aspects—Latin America. 7. Economic development—
Latin America. 8. Social change—Latin America. 9. Latin America—Social conditions.
10. Latin America—Economic conditions.
I. Title.
F1434.2.S62 G63   2001
306'.089'98—dc21   00-064463

No way of thinking or doing, however ancient, can be trusted without proof. What everybody echoes or in silence passes by as true today may turn out to be falsehood tomorrow, mere smoke of opinion, which some had trusted for a cloud that would sprinkle fertilizing rain on their fields. What old people say you cannot do, you try and find that you can. Old deeds for old people, and new deeds for new.

—Thoreau, *Walden*

To Lee, Abipa, Karen, Leandra, and Justin

Contents

Introduction

This book contains three sections and 12 chapters that move the reader from the background and motivation for the study (Part I), to an analysis of how markets have affected the use of natural resources, aspects of social life, and knowledge by indigenous people (Part II), and on to some of the broader conclusions for policy-makers and academics (Part III).

A Road Map

This book begins in chapter 1 by posing the question: "What are the effects of markets on the welfare of lowland indigenous people and the conservation of tropical rain forests?" and discusses the significance of the question in terms of anthropological theory and public policy. The outcomes that dominate Part II are identified and the rationale behind the choice of those outcomes is explained. Chapter 1 also discusses other methods that could have been used to collect information, and discusses problems that were encountered with the research design.

Chapter 2 compares the approaches used by development economists, political economists, and anthropologists in studying the effect of markets on indigenous people. After the comparison, a Ricardian trade model is presented to generate hypotheses about what one might expect to find in the use of natural resources and welfare as markets envelop indigenous people.

Chapter 3 contains a discussion of the weaknesses and strengths of the research methods used are expanded upon and the problems of reverse causality (endogeneity), sample selection, and measurement errors. Chapter 4 provides an ethnographic sketch of the cultures to highlight their long history of contact with the outside world, and the different ways they have used to take part in the market. Chapters 1-4 (or Part I), contain the background in theory, methods, and ethnography for the empirical analysis of Part II.

Part II (chapters 5-12) contains the empirical findings, broken down into topics related to subsistence (chapters 5-7), welfare (chapters 8-10), and cognition (chapters 11-12).

Chapter 5 estimates the effect of markets on the clearance of old-growth rainforest. Through a discussion of theory and empirical evidence, we show that economic development first increases and then decreases deforestation. We show that contrary to considerable anthropological theory, markets do not seem to worsen conservation in a linear way. Rather, markets affect conservation in different ways, depending on the level and type of income. Besides showing the existence of a Kuznets (or inverted U-shaped) curve of deforestation, this chapter explores the extent to which improvements in farm technologies affect conservation and tests whether low rates of private time preference enhance conservation. Chapter 6 presents estimates of the income, cross-price, and own-price elasticity of consumption for game and fish. Edible animals in the rainforest are goods whose consumption increases modestly or declines as incomes rise. The finding suggests that economic development could enhance the conservation of wildlife through the demand side. This chapter presents the results of an animal census done over 2½ continuous years in both a rich and a poor Tawahka village in the rainforest of eastern Honduras to show that economic development seems to have modest effects on the abundance of wildlife.

Chapter 7 tests the idea that demography plays a stronger role in production among relatively autarkic households than among households with tighter links to the market. As economies modernize, demography may wane in importance because consumption and production diverge. Once households have access to well-functioning markets for credit and wage labor, they no longer have to rely on their own laborers to produce what they consume.

Chapters 8-10 examine how markets affect aspects of welfare. American anthropologist Marshall Sahlins' influential idea that leisure declines with modernization was tested, and little theoretical or empirical support for it was found (see chapter 8, "Chayanov and Sahlins on Work and Leisure"). To test the idea, information from scans or spot observations of how people allocate their time is combined with information about their income, wealth, and other socioeconomic and demographic variables. Multivariate techniques are used to estimate the effect of different measures of modernization on different types of leisure.

Chapter 9 explores the effect that markets have on health, defined through self-perceived or through objective criteria. We find that cultural and material determinants both matter (statistically) in affecting health.

Chapter 10 examines the effect of markets on reciprocity, and tests well-established ideas (from Marcel Mauss to the present) that traditional systems of insurance—embodied in gift giving and reciprocal obligations—

break down with modernization, increasing the economic vulnerability of indigenous people. Little evidence is found for the idea that reciprocity weakens with modernization. In fact, the evidence seems to suggest that even in relatively isolated, traditional villages, people do not seem to respond to the mishaps of their neighbors through gift giving as much as we may have thought.

Chapters 11-12 contain analysis of how markets affect aspects of cognition. Chapter 11, written with Nicholas Brokaw, David Wilkie, Daniel Colón, Adam Palermo, Suzanne Lye, and Stanley Wei, estimates how markets erode people's knowledge of forest plants and game. Though much has been written about the loss of knowledge of plants and game, trade theory suggests that the effects of increasing exposure to the market ought to produce different effects on the loss and retention of knowledge. A Ricardian trade model suggests that economic modernization ought to enhance knowledge of forest goods entering commercial channels, but ought to erode knowledge of forest goods replaced by cheaper industrial substitutes. Chapter 11 contains a test of those predictions.

Chapter 12 examines the links between economic development and private time preference or patience. Although private time preference lies at the core of an economy, influencing how much people consume, invest, and save, researchers know relatively little about its socioeconomic determinants or about its environmental consequences in developing countries. Psychologists, economists, and indigenous people themselves have differing explanations for why some people are more patient than others. Chapter 12 tests competing explanations. The results suggest that economic development may attract people with higher private discount rates or those who are more impatient—a process that may accentuate village inequalities, at least in the short run. There seems to be little support for the idea that better human capital lowers private time preference.

In the concluding section (Part III), some of the larger academic questions that motivated the book are answered. Drawing on the information from the societies studied, this section assesses whether economic development increases inequalities, erodes social solidarity, and encourages environmental degradation. Some implications for policy-makers that flow from the analysis are also spelled out.

Style, Audience, and Citations

The book has been written in a simple and didactic style. Econometrics will help in understanding the quantitative analysis, but the logic of the arguments should be accessible to any reader. To keep the book short, citations have been kept to a minimum. To document the intellectual genealogy of each topic discussed would have required writing a larger—though not a more original—book.

Acknowledgments

This book is the result of fieldwork done between 1992 and 1998 by over 20 students and scholars in 65 Amerindian villages and six cultures in the lowland tropical rainforests of Latin America. Such a large undertaking would have been impossible without the financial, logistical, and the intellectual support of many people and institutions.

Financial Support

Financial support for the research came from the National Science Foundation, the Conservation, Food and Health Foundation, the Social Science Research Council, the United States Agency for International Development in Bolivia, and Harvard University.

National Science Foundation

The Cultural Anthropology and Human Dimensions of Global Change programs of the National Science Foundation—through grants SBR-9417570, DBS 9213788, and SBR 9307588—financed fieldwork among the Tawahka of Honduras, the Sumu-Mayagna of Nicaragua, and the Mojeño and Yuracaré of Bolivia. Through its Research Experience for Undergraduates program, the National Science Foundation also paid for the transport and lodging expenses of two undergraduate majors in anthropology, Peter Cahn and Stanley Wei. Cahn and Wei did research among the Tawahka to determine how perceptions of the forest have changed since the early twentieth century (Cahn) and why the Tawahka have lost or retained knowledge of plants and animals from the rainforest (Wei).

Conservation, Food and Health Foundation

The Conservation, Food and Health Foundation financed zoological research among the Tawahka through a grant to Professor Gustavo Cruz of the National University of Honduras' Department of Biology. Through a separate grant, the Conservation, Food and Health Foundation financed research on the effects of education on the use of natural resources by the Mojeño, Yuracaré, Chiquitano, and Tsimane´ of Bolivia.

Social Science Research Council

The Joint Committee on Latin American Studies of the Social Science Research Council and the American Council of Learned Societies financed a study in 1995 on forest clearance among the Tawahka with money provided by the Ford Foundation.

United States Agency for International Development, Bolivia

With financial assistance from the United States Agency for International Development in Bolivia, the Unidad de Análisis de Políticas Sociales (Ministry of Human Development) and the Bolivian Sustainable Forest Management Project (BOLFOR) gave money to two Bolivian (Vianca Aliaga and Julio Romero) and one U.S. (Joel DeCastro) undergraduate students in 1996 to study health, private time preference, tenure security, and forest clearance among the Tsimane´.

Harvard University

The Harvard Institute for International Development, the David Rockefeller Center for Latin American Studies, and the Weitzman Fellowship program awarded research grants to the following students for summer fieldwork among the Tawahka and the Tsimane´: Daniel Colón, Joel DeCastro, Peter Kostishak, Suzanne Lye, Kathleen O'Neill, and Adam Palermo.

Logistical Support and Fieldwork Assistance

Edgardo Benítez, Eusebio Cardona, Dionisio Cruz, Benjamín and Guillermo Dixon, Angel Sánchez, the Federación Indígena Tawahka de Honduras, and Osvaldo Mungia and Suyapa Valle of Mosquitia Pawisa provided logistical and administrative support and fieldwork assistance in Honduras. I would like to thank Evar and Manuel Roca, Zulema Lehm, Carlos Navia, Ramiro Molina, Luis Peñaloza, the Secretaria de Asuntos Etnicos, de Genero y Generacionales, Waldo Tercero of the Proyecto Forestal Chimane, Federico Martínez, Mario Alvarado, members of the Gran Consejo Tsimane', Jorge Añes, Professor Elifredo Zavala, and the village headmen and villagers with whom we worked in Bolivia. Phyllis Glass, Michael Hricz, and Carol Zayotti of the Harvard Institute for International Development provided logistical support to field researchers.

Co-workers

Although I am responsible for the quality of the information collected and for the analysis of the information presented, many students and colleagues helped with the analysis of the information and the writing of results. In addition to the co-authors of some of the sections and chapters, I would like to thank the following people for helping to collect and to analyze the information: Vianca Aliaga, Mario Alvarado, Anupa Bir, David Bravo, Peter Cahn, Marina Cárdenas, Adoni and Glenda Cubas, Josefien Demmer, Verónica Flores, Stephen Groff, Tomás Huanca, Marc Jacobson, Peter Kostishak, Marques Martínez, Josh McDaniel, Han Overman, and Julio Romero. The pronoun *we* in the book refers to the co-authors of some of the sections and chapters, and to the students and colleagues who helped collect and analyze the information.

Comments on Drafts and Discussions

The following people helped to improve the manuscript by commenting on portions of chapters, on entire drafts of chapters, or by discussing some of the ideas presented in the book: Kamal Bawa (University of Massachusetts, Boston), Gary Becker (The University of Chicago), Russell H. Bernard (Florida), Marco Boscolo (Harvard), Michael Chibnik (University of Iowa), Donald Davis (Columbia), Shelton Davis (World Bank), Robert T. Deacon (University of California, Santa Barbara), Mario Defranco (Ministry of the Presidency, Government of Nicaragua), Robert Emerson (University of Florida), Nancy Flowers (City University of New York, Graduate Center), Edward Glaeser (Harvard), Michael Gurven (University of New Mexico), Robert Hunt (Brandeis), Allen Johnson (University of California, Los Angeles), Kris Kirby (Williams), David Laibson (Harvard), Luis Locay (University of Miami), Jonathan Morduch (New York University), Kathleen O'Neill (Cornell), Harry Patrinos (World Bank), Shanti Rabindran (Massachusetts Institute of Technology), Jesse C. Ribot (World Resources Institute), Dani Rodrik (Harvard), Victoria Reyes-García (University of Florida), Alan R. Rogers (University of Utah), Bernardo Rozo (Universidad Mayor de San Andrés), Marianne Schmink (University of Florida), G. Edward Schuh (University of Minnesota), Glenn Stone (Washington University), Cristian Vallejos (Forest Stewardship Council), Bruce Winterhalder (University of North Carolina, Chapel Hill), Richard Zeckhauser (Harvard), the late Michael Roemer (Harvard), and several reviewers for Columbia University Press. Holly Hodder, Lynanne Fowle, and Jonathan Slutsky shepherded the manuscript with patience and good humor.

Much of the ethnographic discussion of private time preference in chapter 12 draws on the observations of Wayne Gill, a missionary who has worked with the Tsimane′ for more than 20 years. Over the years, he has shared his knowledge of the Tsimane′ with me. The observations in chapter 2 about the effects of globalization on local governments come from discussions with G. Edward Schuh.

Last, I would like to thank students at Harvard University, Wellesley College, and the University of Florida for putting up with me when I presented them with the first drafts of the book.

INDIANS, MARKETS, AND RAINFORESTS

Part I

THE QUESTION,
THE RESEARCH DESIGN,
AND THE PEOPLE

Part I, "The Question, the Research Design, and the People," discusses four topics that set the stage for the empirical analysis of Part II.

Chapter 1 discusses why researchers and policy-makers should care about how market economies affect the welfare of indigenous people and the conservation of natural resources in the tropical lowlands of Latin America. The query goes to the heart of fundamental questions about how societies modernize. One can use the question to examine large topics in the social sciences in a quasi-natural laboratory, such as changes in social solidarity or inequality as economies modernize. We may not have many more chances to study such topics because indigenous people are modernizing quickly. On a more practical side, once researchers document aspects of welfare and conservation hurt by markets, policy makers will have better information with which to formulate and carry out public policies.

Chapter 2 reviews methods social scientists have used to study the effects of markets on the culture, society, and environment of indigenous people. Drawing on some of their insights, chapter 2 presents a model that helps explain and predict what may happen to the welfare of lowland indigenous populations as their economy modernizes and the impact on the conservation of tropical rainforests. The model helps explain the empirical findings and some of the ambiguities of Part II.

Chapter 3 discusses the strengths and weaknesses of the method used to collect information, the sampling strategy used to select cultures, villages, and households, and the statistical techniques used to analyze the information.

Chapter 4 presents ethnographic sketches of the cultures analyzed in later chapters. The sketches are short to save space, to stress a few things that matter most, to draw attention to cross-cultural similarities and differences, and to provide a qualitative context for the statistical analysis presented later.

Chapter 1

The Question and Its Significance

The discussion in this chapter tries to accomplish four goals. First, it poses the question, "What are the effects of markets on the welfare of lowland indigenous people and the conservation of tropical rainforests?" and discusses the significance of the question in terms of anthropological theory and public policy.

Second, it identifies the outcomes that dominate the empirical analysis of Part II and explains the rationale behind the choice of outcomes. For reasons of theory, public policy, and personal preference the chapter focuses on how markets shape certain aspects of subsistence (e.g., forest clearance), welfare (e.g., health), and cognition (e.g., private time preference) or, in Marxist parlance, on how markets affect aspects of infrastructure, structure, and superstructure. Taken together, the outcomes provide a wide (though not a comprehensive) picture of what happens to indigenous people and their environment as markets develop.

Third, it justifies the comparative, cross-sectional method and the multivariate techniques used to analyze the information gathered.

Last, problems of research design in the study of how markets affect indigenous people are discussed. The problems include: failure to explain variance in market participation, lax standards in defining variables, failure to examine nonlinear relations between markets and outcomes, excessive reliance on bivariate (rather than multivariate) analysis, failure to capture enough variance in explanatory variables, and insufficient attention to identifying the direction of causality and controlling for unobserved, fixed attributes of people and places.

The Question and Its Significance

Although researchers and policy-makers have been studying and debating the effects of markets on the welfare of indigenous people in the tropical lowlands of Latin America and on their use of natural resources for many years, they can draw few generalizations from the studies and debate. Knowledge of how markets affect welfare and conservation come from ethnographies, but ethnographies cannot be used to generalize because ethnographers do not agree on a common method for collecting, analyzing, and presenting information. Policy-makers and researchers face the proverbial problem of comparing apples and oranges when trying to make sense of the ethnographic record. Over the years, anthropologists have described in detail the stresses (and less often the benefits) produced by the market on the rainforest and on Indian societies of lowland Latin America. They have paid less attention to developing theories and using cross-cultural information to test hypotheses and generalize about the effect of markets on these areas.

Like others before it, this book asks a simple question: "What are the effects of markets on the welfare of indigenous people in the tropical lowlands of Latin America and their use of the rainforests?" Unlike other researchers, however, it answers the question by presenting a theory, deriving hypotheses from the theory, and using comparative, quantitative information from several societies within the rainforests of Central and South America to test the hypotheses. In so doing, the book departs from theoretical improvisation, from reliance on a case study, and from the descriptive and subjective approach that has characterized so much anthropological work in the tropical lowlands of Latin America.

The question of how markets affect the rainforest and the welfare of lowland Amerindians merits attention for at least four reasons. First, one can use the query as a starting point in trying to answer core questions in the social sciences about the evolution of institutions and society. From the writings of nineteenth-century evolutionists to those of the present, researchers have been trying to find regularities in how societies and economies develop. Studying how relatively autarkic indigenous people change in response to greater penetration of markets allows large topics in the social sciences to be addressed in a quasi-natural laboratory. The topics include the evolution of inequalities, social solidarity, and people's subjective valuation of the future or private time preference.

Second, much has been written in cultural anthropology about the contribution of subsistence, social organization, and ideology to behavior. Some anthropologists have said that technology and economics—the material world—shape the way humans think and behave. Others have said that social organization and ideology overshadow material determinants in shaping behavior. A third group has seen two-way causality between ideology and material determinants and has been skeptical about identifying "prime movers." By studying several indigenous societies, each with different degrees of exposure to the market, researchers can estimate the statistical weight of material, social, and ideological determinants and decide which of them matter most as economies modernize. Put differently, material, social, and ideological determinants may carry different statistical weights and may vary in systematic ways as economies modernize.

Third, the topic deserves attention because researchers may not have many more chances to answer the question with primary information (Schemo 1999:72). Indigenous people are now becoming part of market economies and changing their culture at a faster rate. Some topics lend themselves to re-study with tighter methods, better information, and sharper theories than those used by one's predecessors. Other topics do not. As Franz Boas showed in his studies of North American Indians, once relatively isolated indigenous people modernize, researchers forever forfeit their chance to record information and learn how things worked before the great transformation took place.

Last, the study merits attention for reasons of public policy. If markets deepen ecological degradation, increase poverty and economic vulnerability, worsen health, and erode folk knowledge of forest plants and game—as many people say—researchers ought to find ways to contribute to the formulation and implementation of public policies that mitigate the effects. Cultural anthropologists are well positioned to contribute to such a task because of our first-hand knowledge of life in the village. Before formulating such policies, they need to draw on empirical analysis to document how markets may help or hurt welfare and conservation.

Rationale for the Outcomes Examined

By posing the broad question: "What are the effects of markets on the welfare and on the use of rainforests by indigenous people in the tropical lowlands of Latin America?", room is allowed to study how markets affect different aspects of welfare and conservation. In Part II, separate chapters are devoted to how markets affect the clearance of old-growth forest, consumption of wildlife, health, reciprocity, knowledge of forest plants and game, and private time preference.

The outcomes were selected because they matter for conservation, welfare, and for anthropological and economic theory. The net was cast wide to include aspects of subsistence, social organization, and cognition. The sequence of chapters in Part II is not random—it should move the reader from aspects of the material world, to social organization, and on to the world of ideas and cognition.

Other outcomes could have been studied. For instance, topics could have included how markets affect religion, personality, village politics, art, law, language, or child rearing. The topics were left out, important though they may be, because they did not seem to touch on welfare or on conservation as directly as some of the other outcomes chosen.

Rationale for the Cross-Sectional Approach

The most rigorous method for studying the effects of markets on welfare and on conservation consists of giving gifts of cash to villagers selected at random, measuring changes in outcomes before and after the transfer, and comparing changes in outcomes between those who received the transfer (treatment sample) and those who did not (control sample). The experiment allows one to control for reverse causality between the outcome and the explanatory variable of interest (Meyer 1995; Glewwe, Kremer, and Moulin 1998). In studying the effect of markets on conservation and welfare, one is never sure whether markets affect the outcome of interest or whether the outcome of interest affects a person's decision to take part in the market. For instance, taking part in the market may improve health, but only healthy people may decide to join the market. The proposed experiment would break the impasse. In this example, one could not conclude that health affected cash income since the cash transfer would have taken place at random, without taking into account a person's health. Ethical issues aside, randomization would not capture with accuracy the paths and complexity through which markets affect outcomes. Randomization would simply allow one to measure the effect of a windfall gain on selected outcomes.

The second-best method consists in following the same people, households, and villages over time as they take a greater part in the market. The role of unobserved, fixed effects, such as the inclination of some people to become part of the market economy or to drift to the outside world could be controlled or perhaps even eliminated. The relatively pure effect of markets on the outcomes of interest could then be estimated. By using subjects as their own control group, more reliable estimates of parameters could be obtained—but at the cost of lower variance, since variation within a subject is less marked than variation across subjects. Unlike a one-time transfer of cash to people chosen at random, a longitudinal study might be biased by attrition in the sample (Falaris and Peters 1998; Fitzgerald, Gottschalk, and Moffitt 1998). The richest villagers may move out of the village once they pass a threshold of income, leaving researchers with a truncated sample skewed toward the poorest villagers.

The next best method, and the one used most often in this book, relies on a cross-sectional analysis of people, households, and villages at one moment in time. The sample contains people, households, and villages at different points in a rural-to-urban continuum, at different distances from the road or from the market, or with different levels of income. Cross-sectional studies with enough variance in explanatory variables, particularly in degree of exposure to the market, produce reliable estimates of parameters and do not raise the ethical concerns that arise when giving cash to people selected at random. The method has obvious shortcomings. It relies on cross-sectional information to make inferences about a dynamic process, and it cannot control well for endogeneity or for the unobserved fixed attributes of people or localities.

Although anthropologists have criticized the adequacy of using ideal types—such as the folk-to-urban or the autarky-to-market continuum—the criticism is unwarranted in this case. In this book changes between points along the continuum are compared. The generalizations to be drawn as one moves forward or backward in the continuum have significance, but the existence of the end points does not. The use of the continuum does not imply that only movements from left to right or from less to more integration to the market can take place. The theory and the hypotheses presented later also apply when households and societies slide back and move from right to left or from more to less integration to the market, as sometimes happens (Santos et al. 1997).

A New Approach

In examining the effect of markets on the welfare and use of rain forests by Indians, this book draws on—and also departs from—mainstream studies in anthropology. Like other researchers (Baker, Hanna, and Baker 1986; Eggan 1954; Redfield 1941, 1947), this volume compares how different societies have responded to the same market stimuli. Comparative studies of contemporary populations have fallen out of favor in anthropology because anthropologists have grown reluctant to generalize. In the past, anthropologists used the comparative method to infer evolutionary trends. By doing so, they opened themselves to justified criticism about the use of cross-sectional information from contemporary populations to infer changes occurring over a longer period of time. Like other researchers who have used the comparative method, this book stresses regularities emerging from cross-cultural comparisons, but does so without making inferences about evolutionary trends.

Second, the empirical analysis is guided by price and trade theory because together they produce a parsimonious explanation about how markets affect indigenous people. Price and trade theory provide a method for identifying, defining, and measuring variables so one can compare results and make generalizations. Symbolic, historical, and qualitative approaches yield insights into local constraints and mechanisms of change in one culture, but make it hard to compare results and generalize because they anchor information and analysis to one place.

Third, multivariate rather than bivariate analysis is used. Though appropriate as a descriptive first step, a stand-alone bivariate analysis is inadequate for the task at hand because: 1) cultural behavior and biological outcomes generally reflect many causes, 2) multivariate analysis allows one to capture non-linear relations between markets and social and environmental outcomes, and 3) multivariate analysis allows one to control for different scales of analysis (e.g., household, village) at the same time. Only by controlling for the effect of different variables and scales of analysis at the same time can one get an accurate estimate of the extent to which markets—and not close attributes of the market, such as acculturation—affect the outcome. Chapter 9, "Human Health: Does it Worsen with Markets?", contains an example of the limits and biases that may arise from a bivariate analysis.

The Information

To answer the question posed earlier in this chapter, we used similar methods to collect information from several indigenous societies of lowland Latin America: the Tawahka of Honduras, and the Mojeño, Yuracaré, Tsimane´, and Chiquitano of Bolivia.

Those societies do not lie at different points along an idealized autarky-to-market continuum. Rather, the people, the households, and the villages of each society display different amounts of exposure to the market. Households in each society vary in the degree and form of integration to the market they display. For each society, one can estimate how outcomes vary in relation to different degrees and forms of exposure to the market.

The information for these cultures varies in quality and in size (see Chapter 3, "Research Design"). Information for the Yuracaré, Mojeño, and Chiquitano came from ethnographic fieldwork during the summer of 1995 (Mojeño and Yuracaré) and during June 1997–July 1998 (Mojeño, Yuracaré, Chiquitano). Information for the Tawahka came from fieldwork during 2½ years (June 1994-December 1996). Information for the Tsimane´ came from three seasons of fieldwork during the summers of 1995-1997 and during a full year of fieldwork (June 1997–July 1998).

The richest information came from the Tawahka. During 2½ years, five students tracked consumption, use of forest goods, sources and uses of cash, and uses of time of about 300 people in two villages. Though rich in detail, the information for the Tawahka came from a sample of only 32 households. Information for the Tsimane´, Mojeño, Yuracaré, and Chiquitano included more variables and came from a larger sample of households (Yuracaré n=62; Mojeño n=132; Tsimane´ n=209; Chiquitano n=240), but lacked observations on the same households and people over time.

Clearing the Underbrush

Four topics deserve discussion before turning to the theory and to the empirical analysis:

1. the determinants of integration to the market,
2. functional form between markets and outcomes,

3. variance in explanatory variables, and

4. definitions.

The Determinants of Integration to the Market

Anthropologists have described the effect of markets on many outcomes, but they have shown less interest in explaining what lies behind variation in integration to the market. One cannot take different degrees of exposure to the market as given. For reasons of ethnography and method, one needs to explain why some villages, households, and people are more linked to the market than others.

Many anthropologists have said that unacculturated indigenous people enjoy better health, more leisure, and higher (and more stable) consumption than more acculturated people. If that is true, one needs to explain why people in the "original affluent society" (Sahlins 1968) would leave autarky to take part in the market. If autarky contains so many benefits, why do people in households, households in villages, and villages in ethnic groups display different degrees of participation in the market?

Integration to the market—unlike variability in rainfall—is not an exogenous event that randomly affects some people and households more than others. People and households self select how much of the outside culture they will absorb, how much they will sell or buy in the market, and how much credit they will receive from moneylenders. Extreme forms of exploitation aside (MacLeod 1979), indigenous people are not supine victims of markets. Instead, they select the form and the amount of integration to the market they wish to achieve. The existence of choice makes it more difficult to obtain accurate estimates of how markets affect welfare. If indigenous people simultaneously select the amount of integration into the market they want to achieve and, say, the amount of modern health care they want to buy and researchers do not control for the simultaneity of the decisions, they will get flawed readings of how markets affect welfare.

Few anthropologists have wrestled—in a rigorous, quantitative way—with the source of variation in exposure to the market. The neglect may stem from several reasons. First, measuring the source of variation of an explanatory variable could lead to an infinite regress. For instance, differences in integration to the market could reflect differences in pressure by outside encroachers. Encroachment by outsiders, in turn, could reflect discriminatory government policies against the countryside and rural poverty.

Discriminatory government policies could reflect urban bias, and poverty could reflect low levels of schooling, and so on. To set analytical boundaries and make estimation possible, researchers must assume that some variables are fixed or given.

Second, variation in exposure to the market generally comes from processes that originate outside the village. Since anthropologists work chiefly in villages, they sometimes miss the forces beyond the village that push villages and households toward or away from the market. It is true that anthropologists often discuss the links between different levels of analysis or social organization and provide qualitative descriptions of those links (DeWalt and Pelto 1985; Morán 1993a, 1993b). They have not, generally, provided quantitative estimates of how different levels of social organization affect the behavior of rural people. The neglect probably reflects the costs of sampling. To estimate the effect of one village variable (e.g., presence of a health worker) in a cross-sectional study, one needs a sample of at least two villages; to estimate the effect of one regional variable (e.g., a strong form of municipal government), one needs a sample of at least two regions; to estimate the effect of one national-level variable (e.g., exchange rate), one needs a sample of at least two nations. Few anthropologists have the money to carry out quantitative research in several regions or countries. Much of the discussion about how different levels of social organization affect behavior, therefore, remains qualitative.

Third, one must also measure or control for unobserved, fixed attributes of people or localities to understand why some people or households take a greater part in the market than others. Some indigenous people show more interest in learning about a foreign culture or becoming a member of a foreign group than others (Murphy 1961-1962). Even if outside encroachers do not push indigenous people to the market, some indigenous people will take a greater part in the market than others by personal preference. Because our tools for measuring unobserved cognitive variables are crude, we generally leave out of the analysis how personal preferences and other fixed unobservables affect exposure to the market.

Functional Form Between Markets and Outcomes

Anthropologists often assume that markets affect welfare and conservation in a linear way. Many seem reluctant to entertain the possibility that beyond a threshold of economic development or of income things may change. They do not entertain the idea that the quality of the environment may get worse before getting better, or that it may get better before getting worse. The relation between markets, welfare, and conservation may not be as linear as researchers have assumed.

As seen later, there are theoretical reasons to believe that markets exert non-linear effects on many outcomes. For instance, low levels of participation in the market seem to increase deforestation, but higher levels of participation in the market seem to reduce deforestation even after controlling for the amount of forest available.

Variance in Explanatory Variables

Researchers have seen a linear relation between markets and welfare or conservation, and this, in part, stems from the limitations imposed by traditional methods of anthropological fieldwork on the information collected. Anthropologists studying the effects of markets on indigenous people have typically carried out their fieldwork in villages. Indigenous people who have passed a threshold of integration to the market or income may move outside of the village and of the group observed by the researcher. Indigenous people who have moved out of the village forage and farm less, work in non-farm occupations, and often enjoy access to modern health-care facilities. Unless anthropologists capture the right-hand tail and middle of the distribution of income, exposure to the market, or acculturation, they will end up working with truncated information. With truncated information, researchers find it difficult to see parabolic relations between participation in the market and social or environmental outcomes.

Definitions

Much effort was put into defining variables with care—for two reasons. First, results of analysis may be sensitive to how one defines variables. Second, failure to define variables with care could bias estimates.

In Part II, several examples of how results change depending on how one defines explanatory variables are presented. For instance, in Chapter 11, "Of Trade and Cognition: On the Growth and Loss of Knowledge," we show how the retention and loss of knowledge of forest plants and game varies in systematic ways with different types of integration to the market. Taking part in the market by working for a wage produces a different effect on knowledge of flora and fauna than taking part in the market by selling crops. In Chapter 5, "Forest Clearance: Income, Technology, and Private Time Preference," we show that working for a wage reduces forest clearance, but that producing annual crops increases forest clearance. Researchers need to ensure that conclusions hold up irrespective of the definitions used.

Failure to define outcome variables with care can lead to misleading conclusions. Different stories of how markets affect outcomes can be told, depending on how one chooses to define outcome variables. For instance, integration to the market could either improve or worsen health, depending on whether one defines health through objective or self-perceived criteria (see Chapter 9, "Human Health: Does it Worsen with Markets?").

Clear definitions are necessary to avoid biasing estimates. The determinant of most interest to us—integration to the market—overlaps with other variables, such as modernization, acculturation, and urbanization. Indigenous people or households that take a greater part in the market also tend to be more acculturated, own more modern physical assets, and live in towns or crowded villages. Unless researchers measure and control for the close covariates of the market, they will arrive at inaccurate estimates of how markets affect indigenous people. The estimate will be biased because the variable for integration to the market will pick up some of the credit or bear some of the costs of the variables that overlap with the market but that researchers leave out of the analysis.

A theoretical example will clarify what I mean. Suppose indigenous people who take part in markets have better health than indigenous people in relative autarky, but suppose that the people who take part in the market have also absorbed more of the culture of outsiders. Suppose, further, that acculturation bears a positive link to the market, with more acculturated people also selling and buying more labor and goods than less acculturated people. If researchers estimate the effect of markets on health without, at the same time, measuring acculturation they will conclude that markets improve health. In fact, some of the benefits we would be attributing to the market would be coming from acculturation rather than from the market. Chapter 9, "Human Health: Does it Worsen with Markets?" contains a real variant of this example.

This book is about how markets affect the welfare and the environment of any rural society. Although the indigenous people of the tropical rainforests of Latin America are the societies used to develop and to test hypotheses, neither the use of indigenous people nor the use of the Neotropics are necessary to test the hypotheses. The theory and the hypotheses apply to any rural population in any place and could have been tested with any other rural population undergoing incorporation into the market.

In this book, a naive approach was taken in defining the word indigenous, allowing people to decide for themselves whether or not they considered themselves part of the ethnic group. The choice of tropical rainforests to examine the effect of markets, though unimportant from the viewpoint of the theory and of the hypotheses presented, is nonetheless important from the point of view of public policy. Because tropical rainforests contain most of the world's biological diversity, that region is a good site for study if one wishes to contribute to a prominent public policy debate.

Although the central question raised in this book has been asked before, the approach taken here to answer the question is new in several ways. The approach is comparative, quantitative, and rests on price fad trade theory, and a clear definition of terms.

Chapter 2

Comparing Approaches

As in other parts of the developing world, the expansion of markets in the tropical lowlands of Latin America has often followed conquest and migration. Indigenous people in the lowlands have joined the market as part of business cycles of forest goods, or as part of a longer and less reversible trend. Something as complex as the incorporation of indigenous people into a market economy—with so many regional variants and with so many historical nuances—requires the use of different theories and disciplines to be understood well. No approach can explain in full the many causes and consequences of participation in the market across all cultures.

This chapter compares the approaches used by development economists, political economists, and cultural anthropologists to explain what drives indigenous people in the tropical lowlands of Latin America to take part in the market—and the socioeconomic and environmental consequences of such a process. The same process—greater participation in the market—looks different depending on the academic lenses. Different disciplines differ in the assumptions they make and in what they can and want to see. The comparison allows one to assess the strengths and weaknesses of different approaches, identify gaps, and cull the best of each.

The approach of development economists, political economists, and cultural anthropologists overlap, making it hard to identify where one approach ends and where another one begins. Many authors live amphibiously in two camps. Some topics are analyzed by all groups, but only underlined by some. Development and political economists, for example, recognize the role that discriminatory government policies against the countryside plays in driving poor rural producers to encroach on indigenous territories and driving indigenous people into the market. Political economists stress the history of discrimination, whereas development economists stress the economic consequences of discrimination. The discussion below highlights differences between approaches, putting aside areas of overlap. After reviewing the three approaches are reviewed, the intuition behind a Ricardian trade model and several hypotheses that flow from the model are presented.

The Approach of Development Economists

Economists have not discussed what happens to the welfare of indigenous people or to their use of natural resources as market economies expand, but one can draw on development economics for some guidance.

The record of Western (and some Asian) nations suggests that in poor countries, or those in the early stages of economic growth, agriculture absorbs most of a country's land, capital, and labor. As agriculture modernizes, it becomes an engine of general economic growth—pushing rural people out of the countryside and luring rural people into higher-paying urban jobs by encouraging growth in activities unrelated to agriculture. Through technological innovations in agriculture, which include the development of new plant varieties, new practices for tilling the land, new forms of storing and of transporting food, new arrangements over property rights, and through public investments in the countryside (Alston, Libecap, and Schneider 1996), the productivity of land and labor rises. As the national economy develops, the contribution of agriculture declines, overshadowed by the growth of services and manufacturing (Schuh and Brandão 1992).

In analyzing the movement of rural people out of agriculture, one should distinguish between the migration out of the countryside due to the modernization of agriculture, and the migration out of the countryside due to government policies that discriminate against rural people. The import-substitution policies of Latin America before the 1980s favored urban consumers at the expense of rural producers, lowering rural income, and driving rural people into cities (De Janvry 1981).

The mechanisms by which the modernization of agriculture fuel rural-to-urban migration are not easy or automatic (Timmer 1995). Low prices of food in the world market (caused in part by dumping and subsidies of industrial countries), price volatility of crops, and the long-term decline in the inflation-adjusted price of food in world markets all combine to send ambiguous signals to policy-makers about the importance of agriculture in national development. Western countries in the past, and some Asian countries at present, overcame the hurdles by investing in agricultural research and extension, physical infrastructure, schooling, and schemes to stabilize prices. In much of Africa and Latin America, biases against agriculture persist in the form of overvalued exchange rates and low investments in agricultural research, extension, physical infrastructure, and rural education (Bautista and Valdés 1993; Schiff and Valdés 1992).

Why does labor move out of the countryside during economic development and what are the implications of this type of migration for indigenous people and for conservation in the tropical lowlands of Latin America? Discriminatory government policies aside, labor leaves the countryside because of technological transformations in agriculture, which affects the output and price of staple crops (Lipton and Longhurst 1989). Breakthroughs in the technology for producing relatively non-traded staples (or crops that generally do not enter the world market) cause the price of staples to fall because producers cannot export the increased output. The fall in price benefits many people outside of agriculture because of the multiple links between agriculture and the rest of the economy (Bell, Hazell, and Slade 1982; Haggblade, Hazell, and Brown 1989; Hazell and Ramsamy 1991; Johnston and Kilby 1975; Mellor 1976). In Malaysia and India, for instance, one dollar produced in the countryside from public investments in agriculture generated an additional 80 cents in the rest of the economy from greater household expenditures (Bell, Hazell, and Slade 1982; Hazell and Ramsamy 1991).

Lower food prices increase real urban wages, while nominal wages remain constant. Since poor people in cities spend a larger share of their income buying food than the wealthy, lower food prices improve the income distribution of the country (Mellor 1988; Scobie and Posada 1978). Early adopters of new technologies—typically larger farmers—see their incomes rise with the introduction of new farm technologies. If they have secure property rights to land, early adopters increase investments in their farms. If

they do not have secure property rights to land, higher income makes it easier for early adopters to demand titles or to accentuate private rights of use
or ownership to their land. Rural producers who cannot adopt the new
technologies—typically the smaller farmers or those living in areas unsuited
for the adoption of new plant varieties—see their farm income drop when
the price of staple crops falls. Adopters buy the assets of non-adopters and
help lubricate the movement of non-adopters out of agriculture. After
receiving payment for their land and assets, non-adopters migrate to cities
or stay in the countryside to work as laborers for those who adopted the
new technologies.

The evidence from southern India suggests that winners in the countryside during the agricultural transformation included adopters of improved
plant varieties and non-adopters with small plots. The farm income of non-
adopters fell as the agricultural transformation unfolded, but their wage
income rose because those who could not adopt began working as wage
laborers for those who did adopt. Only laborers without land were worse
off after the introduction of new plant varieties than before (Hazell and
Ramasamy 1991). Technological innovations in agriculture, at least in
India, seemed to produce more winners than losers—even in the countryside
(Jirström 1996; Tripp 1996).

But the benefits of the agricultural transformation do not stop in the
countryside. The effects reach beyond the countryside into cities and industries. Breakthroughs in production technology for staple crops reduce the
costs of producing manufactured goods and services, and fuel the expansion
of activities unrelated to agriculture by raising real urban wages while keeping nominal wages constant. On the demand side, the extra available
income gained from paying lower food prices increases the demand for
many goods and services outside of agriculture. On the supply side, the
decreased cost of production from lower food prices causes an increase in
activities unrelated to agriculture. As employment opportunities grow in
urban manufacturing and services, rural people displaced by the agricultural
transformation find work in non-farm occupations. For instance, improvements in the technology for production potatoes in Bolivia (one of the country's core staple crops) could reduce national unemployment by 10% (De
Franco and Godoy 1993:574). Greater output from industry and agriculture
increases national savings and government tax revenues, which governments
can use to finance economic growth. The decrease in the wage costs from
the decline in food prices allows industries and services outside of agricul-

ture to compete in international markets. When agriculture modernizes from technological breakthroughs, the countryside produces more food than it needs—the surplus becomes a fund for investment in activities outside of agriculture.

The flow of labor out of agriculture and the production of agricultural goods in fewer hands makes it easier for rural producers to organize and pressure the government for greater and better public investments in such things as agricultural research and extension, physical infrastructure, and human capital. In the early stages of economic development, governments discriminate against agriculture and treat rural areas as a source of economic rents. As economies develop, governments end up protecting a smaller but increasingly influential lobbying group of rural producers (Lindert 1991).

Many researchers have found evidence of growing income disparities in villages as the agricultural transformation unfolds (Griffin 1979; Vandana 1992). The diagnosis is correct, but incomplete when seen from the Olympian viewpoint of the entire economy. Landless rural people pushed to the cities may lose as producers, but they gain as consumers. The agricultural transformation leads to a more progressive distribution of income when urban labor markets work well and when markets can absorb workers leaving the countryside. Otherwise, rural people displaced by the agricultural transformation stay trapped in agriculture, skewing the distribution of income even more.

The above approach fits the experience of Western (and some Asian) nations better than it does the experience of African or Latin American nations. Despite its limitations, the approach has several strengths. The approach is explicit about how it operationalizes integration to the market. Integration to the market is equated with monetary income. What is more important, the approach helps to generate hypotheses about what might happen to indigenous people during economic development.

First, the model predicts that economic development will push and pull indigenous people out of rural areas. Second, the model predicts that those who stay behind will be either the wealthy adopters or those who cannot move into industrial jobs because they lack the skills to compete with success in modern labor markets. Third, the model predicts that income inequalities in villages and ethnic groups may increase in the short run. Adopters may be pitted against non-adopters. Over time, however, income inequalities will decline if markets work well.

The economic model assumes that greater participation in the market economy will raise income (Huber 1971), consumption, education, and the quality of leisure. In recent years, some economists have said that the traditional insurance schemes of rural economies—such as sharecropping and gift giving—may break down as rural economies modernize, increasing the economic vulnerability of the rural poor. Researchers still have not yet measured well the extent to which modern institutions, such as credit markets, replace traditional forms of insurance (see Chapter 10, "Mishaps, Savings, and Reciprocity").

The approach assumes indigenous people will migrate or stay behind—will sell or buy—prompted by movement in relative prices brought about by breakthroughs in the technology for producing staple crops. The approach assumes indigenous people will act to increase their individual welfare, but ignores political forces pushing indigenous people to take part in the market against their will. Political economists address that challenge.

The Approach of Political Economists

Researchers who take a political economy approach come from anthropology (Collins 1986; Davis 1977; Painter and Durham 1995; Schmink and Wood 1987, 1992), geography (Hecht 1985, 1993), political science (Foweraker 1981), and sociology (Bunker 1985). Though they come from different disciplines, they draw on political science and economics to examine how government policies affect the apportionment of economic rents and the consequences of rent seeking for indigenous people. Researchers who take a political economy approach have done most of their work in Latin America (particularly Brazil), though some have written about Asia and Africa (Ribot 1993; Watts 1987; Wilmsen 1989). The approach peaked in the mid 1980s, when macroeconomic chaos ravaged Latin American societies (Hurrell 1990:198).

Political economists start by explaining the historical and social causes that produced the rush to colonize the Latin American lowlands during the 1960s-1970s (Bunker 1985; Hecht and Cockburn 1989). During that time, private investors started to hoard land in the tropics of Latin America, prompted by government incentives, macroeconomic chaos, and by the shortage of savings institutions. Government incentives included tax holidays, subsidies, and tax credits for corporate investors (Hurrell 1990:202-

203). The incentives made it attractive for investors to clear forest for cattle or cash crops. High inflation and a shortage of modern savings institutions encouraged cattle ranchers, commercial farmers, and small farmers to claim and hoard physical assets against the declining nominal value of money. Muddled property rights accentuated the race to clear forest and claim land before others stepped in.

Political economists stress the web of government policies, macroeconomic chaos, and institutions that lured firms to move into the tropical lowlands. They also stress the causes that pushed poor small farmers into the same area to eke out a living. Persistent discriminatory policies against rural areas lowered incomes and pushed the poor to mine the resources of the land, migrate into marginal areas of the countryside, and reduce investments in conservation (Collins 1986; Zimmerer 1993). The movement of firms and small farmers into the tropical lowlands pushed indigenous people into marginal areas and increased conflicts between indigenous people and abutters. It may also have increased the depletion of natural resources by indigenous people themselves if they perceived greater insecurity of tenure over their natural resources (Godoy et al. 1998, 2001).

Though political economists put the blame for environmental degradation and poverty on the shoulders of the state, they also recognize that the state was not a pawn of international capital or the elite. The state embodied conflicting interests and acted on its own, often independent of civil society (Cleary 1990; Hurrell 1990:203).

The orthodox political economy approach has changed in recent years to acknowledge the growing role of local governments—and the shrinking role of the central government—in shaping the use of natural resources by villagers. The change in emphasis has resulted for at least three reasons. First, national policies have become less relevant in accounting for the use of natural resources. Before the 1980s, state policies helped to push poor people into the forest, but state policies became less relevant in explaining the use of natural resources once poor people had moved into the hinterland. The removal of many large economic distortions in Latin America after the mid-1980s has not curbed environmental degradation by smallholders, suggesting that local determinants must be at work.

Second, the increased openness of national economies has weakened the influence of the central government in some areas. As a result of a more open economy and political system, parts of policy-making and policy implementation have moved to the international level and have become embedded in new international institutions and arrangements. International bodies have taken over some of the areas that the state used to handle, such as the environment, human rights, and trade (Hurrell 1990:211).

Last, to achieve greater efficiency and accountability, policy-makers in Latin America in the early 1990s started to devolve administrative and budgetary responsibilities to local governments (Nickson 1995). As decentralization expanded, the central government became less relevant in local decisions about how to use natural resources.

For all these reasons, local governments have gained prominence relative to the central government in decisions about how to allocate public expenditures and use natural resources. As local governments gained power, political economists have shifted their gaze to middle levels of social organization. By focusing on municipalities, counties, or departments rather than on national policies, political economists have started to explore national heterogeneity in the use of natural resources—a topic de-emphasized in the old school of political economy.

It is too early to assess the usefulness of the new approach. On the methodological front, researchers have made advances by combining satellite imagery on land use with socioeconomic and demographic information from surveys of units such as municipalities. The synthesis allows one to document changes in land use over time. On the policy front, the new approach promises to yield information that will help tailor government policies to the needs of different regions.

But the new approach faces challenges. First, the use of aggregate information for different regions makes it hard to link geographic units with decision units (King 1997; Wood and Skole 1997). Municipalities do not decide how much forest to cut—people, firms, and households do. Second, the use of secondary information (whether from census or from satellites), does not allow researchers to test novel hypotheses that require the use of primary information. Researchers are stuck with information produced for purposes other than the study of natural resources or rural populations. Third, the use of aggregate information does not allow researchers to explain the mechanisms through which explanatory variables affect environmental outcomes. Faced with these constraints, some researchers have started to supplement analysis of secondary information from municipalities with primary information from case studies (Hecht 1998; Wood and Skole 1997).

What do political economists say about why indigenous people become part of markets and about the consequences of taking part in the market?

Political economists assume that the move to the market by indigenous people reflects neglect and discrimination against the countryside by the government and encroachment by those with power or by poor smallholders who move into the territory of indigenous people. Encroachment by outsiders pushes indigenous people to the market. Political economists assume that greater integration into the market lowers income and increases consumption of commercial goods, undermines nutrition and health, reduces leisure, and degrades the environment—unless communities control their resources.

Political economists have described how different levels of policy-making link with one another, and how different types of variables—macroeconomic, political, and social—influence the use of natural resources and social equity. Though rich in description, the approach of political economists has not produced many cross-cultural generalizations. A few anthropologists have taken up that challenge.

The Anthropological Approach

The two approaches discussed so far highlight the tension between the wish to generalize and the wish to acknowledge specifics, between micro and macro levels of analysis, between internal and external determinants to the system, between description and analysis, and between analysis and policy-making. In the late 1970s, anthropologist Daniel Gross and his co-workers attempted to reconcile some of these tensions in a study of how markets and acculturation affected the quality of life and natural resources of indigenous people in central Brazil (Gross et al. 1979). Their article, *Ecology and Acculturation Among Native Peoples of Central Brazil*, deserves close scrutiny because it contains large ideas, because it continues to stir debate two decades after its publication, and because it can teach one a great deal for future research (Santos et al. 1997).

The Model of Gross and Colleagues

In their article, Gross and his collaborators tried to explain why indigenous groups (rather than individuals or households) differ in the amount of time they spend in commercial activities. They start the article by saying that researchers in the past have said that the lure of the market explains why indigenous people participate in the market. Contact between an indigenous and a modern society is enough to make indigenous people want industrial goods. To get those goods, indigenous people must work for a wage, sell goods, or do both. They call this line of reasoning the *culturalist hypothesis*.

Gross and his co-workers reject the culturalist hypothesis and advance their own explanation that draws on political economy and on cultural ecology. They say that circumscription or encroachment by outsiders reduces the amount of land available to indigenous people for subsistence. With less land and increasing population, indigenous people intensify production by reducing the length of fallow, but in so doing they degrade the environment. As land degrades, indigenous people join the market to make ends meet. Once started, integration to the market and ecological degradation feed on each other. People trade for industrial goods so they can prevent farm productivity from falling, but new technologies accelerate "habitat degradation and the process of dependence on the market to meet needs" (Gross et al. 1979:1049).

To test the two hypotheses, Gross and his co-workers measured the allocation of time to subsistence and market activities, crop yields, length of fallow, intensity of cultivation, time since first contact with Western cultures, and knowledge of Portuguese in four Indian societies of central Brazil—Xavánte, Kanela, Bororo, and Mekranoti—during 1976-1977. Table 2-1 contains a summary of information on the four societies, with time allocated to market activities expressed as a share of total time devoted to work rather than in the amount of hours spent in market activities.

Time allocated to market activities is expressed as a share, or as the amount of time an average person devotes to market activities divided by the total amount of time that person devotes to all work (subsistence plus market activities). Gross and his co-workers used the amount of time a person devoted to market activities rather than shares. The two methods produce different rankings of integration to the market. For example, the average Kanela spent a total of 542 hours in market activities. Gross and his colleagues considered the Kanela to be the culture most heavily linked to the market. The Kanela, however, also spent many more hours working in both subsistence and in market activities than any other group. The average adult spent 1,723 hours per year working. When expressed as a share of total work effort, the Kanela spent less time in the market than the Bororo.

TABLE 2-1 Effects of Markets on Conservation in Central Brazil

Ethnic Group	Acculturation				Market	
	Km.	Years	Portuguese	Labor	Fallow	Yields/ha.
Mekranoti	500	10	1%	15%	n/a	29
Xavánte	35	25	2%	19%	20	8
Kanela	35	160	39%	31%	12	16
Bororo	20	85	98%	33%	10	5

SOURCE Gross et al. (1979)

NOTES km. is distance to nearest settlement where trade is possible.
Years is time since first contact with Western culture.
Portuguese is share of population speaking Portuguese.
Labor is share of time allocated to market activities.
Fallow is years land lies fallow.
Yields/hectare is kilocalories x 10^6 per hectare for maize, beans, rice, manioc, yams, and sweet potatoes.

Labor devoted to market activities is expressed as a share rather than in levels for didactic reasons raised in the last chapter—how one measures integration to the market may matter in the results. As we shall see, using this definition, it becomes harder to reject the culturalist hypothesis dismissed by Gross and his colleagues.

Despite its merits (to be discussed shortly), the study contains several shortcomings in methods and interpretation that one needs to discuss for the benefit of future researchers. First, Gross and his colleagues measured integration to the market in only one way—through the number of hours devoted to market activities. To ensure the results of their study were robust to definitions, they should have also used: 1) total cash income from the sale of food crops, Brazil nuts, wage labor, or fish, and 2) consumption and expenditures on goods from the market.

Second, the authors dismissed the culturalist hypothesis even though the information they presented suggested that it had merit. To test the culturalist hypothesis that integration to the market resulted from simple exposure to Western goods and ideas, Gross and his colleagues collected information on two variables: 1) years elapsed since first contact, and 2) share of the population speaking Portuguese. The authors concluded that length of contact with the outside culture bore little relation to degree of participation in the market: "the Xavánte and Mekranoti have been in contact for about the same length of time, yet there is a *wide discrepancy* in their degrees of market involvement" (Gross et al. 1979:1048; my emphasis).

But another interpretation is possible. If one regroups the information from table 2-1, with length of contact on the x-axis and share (rather than levels) of time devoted to market activities on the y-axis, one sees a positive relation between the two variables with a slight dip toward the right-hand side of the curve. The conclusion that there is no relationship between length of contact and participation in the market is too strong given the evidence presented.

Since they also collected information on the share of the population that knew Portuguese, they were able to test the culturalist hypothesis by estimating the relationship between time allotted to market activities and knowledge of Portuguese. Though they do not carry out the test, the authors dismiss the idea that knowledge of Portuguese could increase the desire for Western goods (Gross et al. 1979:1048). Again, if one regroups the information from table 2-1, with share of people knowing Portuguese on the x-axis and share of time allotted to market activities on the y-axis, one sees a positive relationship between the two variables. People in communities with a high percent of Portuguese speakers seemed to spend a greater share of their time in activities related to the market.

Could integration to the market result from distance to the nearest market town? The authors say no—the Kanela "are about as isolated from Brazilian settlements as the Xavánte and still they spend about 275 hours more per year on market activities" (Gross et al. 1979:1048).

Their information shows that the Mekranoti, who lived 500 kilometers from the nearest road (the most remote community in their sample), spent the least amount of time on market activities (15%). At the other extreme, the Bororo, who lived closest to the market town, spent the greatest share of the time in commercial activities (33%). Based on their four communities, one could conclude that as distance to the nearest market town increased, the share of time allotted to market activities decreased. The Kanela and Xavánte, both 35 kilometers from the nearest towns, were above and below a downward sloping line.

Third, Gross and his colleagues did not collect direct measures of circumscription—their central explanatory variable. Direct measures of circumscription would have included the number of abutters next to villages or the number and intensity of conflicts with abutters. Instead, they measured circumscription indirectly by using agricultural intensification as a proxy. They used a summary variable (S) to measure pressure on resources. The S variable took into account the ratio of crop yield to labor input, dependency ratio of the community, and the area of new land cleared every year. Time allotted to market activities and S bore a positive relation to each other, as Gross and his colleagues predicted. The S variable received much attention in their article, but it remains an imperfect, indirect proxy of circumscription. Internal population growth without circumscription could produce intensification or a high S.

Fourth, Gross and his colleagues were aware of potential problems with endogeneity, but did not identify the direction of causality. Although integration to the market could produce ecological degradation, ecological degradation induced by internal population growth (with no encroachment) could have pushed people to the market.

Fifth, the analysis suffers from unnecessary aggregation of information. All the relations discussed by Gross and his colleagues refer to averages for ethnic groups, but some of the information they collected came from individuals. Many of the links they tested for communities could have been tested with a larger sample had they used information from individuals. For instance, they could have tested whether knowledge of Portuguese affected the amount of time people spent in market activities by carrying out a bivariate analysis of individuals' Portuguese knowledge (explanatory variable) against time allotted to market activities (dependent variable). Instead, they aggregated information to communities and thus reduced the sample size to only four observations.

Last, Gross and his colleagues were unable to estimate the separate and simultaneous effect of circumscription and acculturation on indigenous people's decision to become part of the market. They provided a bivariate analysis, which leaves much room for discussion about the role of omitted variables, levels of statistical significance, and interaction between explanatory variables such as acculturation and circumscription.

Despite flaws in the method of analysis and in the interpretation of the information, Gross and his colleagues raised questions that remain unanswered to this day. First, they posed a query of fundamental importance: "What drives indigenous people to take part in the market?" Since the article appeared, we have learned that the forces that drive indigenous people to increase their participation in the market include themes stressed by political economists (e.g., government policies) (Santos et al. 1997), the wish to increase consumption and income, increased sedentariness, and competition over scarce resources in areas with diverse habitats (Cashdan 1987). Participation in the market can reflect multiple and simultaneous attributes of the person, household, village, and region. Within an indigenous group, some people drift to the market from personal preferences, wealth, village proximity to the road, or from circumscription. Even though some explanatory variables may be more important than others in a statistical or social sense, many of the determinants operate at the same time, often interacting with each other.

Second, the article set a tone about the harmful effects of markets on the welfare and on the environment of indigenous people that lingers to this day in the writings of many anthropologists, but it did so in an empirical fashion rather than in a polemical fashion. The authors presented quantitative evidence showing that environmental degradation and agricultural intensification seemed to go hand-in-hand. They presented the information they used in their empirical analysis in a clear, succinct way, allowing other researchers to re-analyze the information and arrive at different conclusions.

Third, the analysis raises questions about sample size, aggregation of information, selectivity bias, the need to control and measure for both acculturation and markets at the same time, and the need for variance in explanatory variables (Godoy, Wilkie, and Franks 1997). More importantly, a close reading of the article reveals the difficulty researchers face in examining the effect of markets on the welfare and environment of indigenous people.

The article by Gross and his colleagues contains adequate empirical analysis and useful insights, but it lacks a theory to generate hypotheses. As we shall see, encroachers do not need to drive indigeneous people to trade. Trade confers many advantages on indigenous people and produces good and bad effects on welfare and on the environment. The next section contains the intuition behind a Ricardian model of trade, and an explanation of why indigenous people may enter the market in a voluntary way, and a discussion of the consequences of such a move on their welfare and their environment.

A Ricardian Model of Trade

We start with two idealized poles—autarky and market—at each end of a continuum. Indigenous societies at the autarkic end rely on hunting, gathering, horticulture, and reciprocity. Households produce all or most of what they consume (Locay 1990). Societies at the other end rely on well-functioning markets for goods, capital, and labor. Households in more developed economies produce little of what they consume, relying, instead, on specialists. In practice, each economy along the continuum contains traits of both poles and may slide back. The end points are ideal types; the stress in the discussion below is on generating hypotheses about what happens as one moves along the continuum in either direction rather than the empirical existence of the end points. The model helps to predict what happens when people become more or less integrated to the market.

With the continuum set, we now draw on a Ricardian trade model to examine the effect of markets on:

- the use of natural resources
- subsistence
- forest clearance
- economic specialization
- reciprocity
- leisure
- knowledge of plants and wildlife

Many of the points outlined below are discussed at greater length and tested later.

As indigenous economies open to trade with the outside world, villagers begin to export goods to outsiders from the village common (e.g., timber). As exports expand, the quantity of goods extracted increases and the quantity consumed in the village decreases. With trade, villagers who specialize in extracting goods for export gain while villagers who only consume the goods lose from higher prices and reduced amounts left over for local consumption.

It follows that when villages open up to trade with the outside world, the availability of forest goods for export declines relative to the availability of forest goods for consumption. As the price of forest goods in demand by outsiders rises, people start to specialize in collecting those goods.

Changes in the value of people's time reinforce the shift to extract forest goods of higher value. When relatively close indigenous economies open up to trade, the value of time for rural people rises because they have new employment opportunities outside of the village. Faced with more employment opportunities, foragers specialize in extracting goods of higher value because only those goods can compensate them for the higher value of their time. Two forces—the rise in the relative price of goods for export and the rise in the value of people's time—induces villagers to specialize in extracting only a few goods from the forest, as Murphy and Steward (1956) taught long ago.

With trade, however, commercial goods start to enter the village economy and start to displace traditional goods and crafts. Tin roofs and manufactured clothing replace thatch roofs and bark clothing because they last longer and require less time to make. Pills replace many traditional medicines because pills cost less. Guns replace bows, arrows, and blowguns because they last longer and are more efficient at capturing animal proteins (Hill and Hawkes 1983; Vickers 1994:320; Winterhalder and Lu 1997:1361; Yost and Kelley 1983). Cloth bags replace hand-woven bags (Wagley 1955:105). With the expansion of trade, the extractors of traditional forest goods and traditional craftsmen suffer because the goods they used to extract, barter, and make are replaced by cheaper goods from the outside world. The availability in nature of forest goods replaced by industrial substitutes increases relative to the availability in nature of forest goods exported from the village.

Trade with the outside world can improve or worsen conservation, depending on the type of forest good traded. Trade ought to reduce the availability in the forest of forest goods exported from the village, but trade should increase the availability in the forest of forest goods replaced by imported goods.

Many ethnographic and empirical observations begin to fall in place from this simple line of reasoning. If trade induces people to specialize in extracting a few goods from the forest, it follows they will know more about plants and wildlife entering commercial channels than about plants and wildlife displaced by substitutes. Greater integration into a market economy ought to produce greater knowledge of fewer forest goods and should not erode knowledge of flora and fauna in a uniform way. For instance, where eco-tourism puts a premium on knowing about charismatic megafauna, villagers should end up knowing more about the nesting and feeding habits of selected animals in the forest than they did before trade with outsiders intensified. Specialization from trade points to where people will gain or lose knowledge.

Markets will also affect the allocation of resources in the village economy. With greater integration into the market, the price of tradable forest goods increases relative to the price of non-tradable staples. This should drive indigenous people to shift to forest-centered activities of higher value, such as ecotourism or extraction of valuable timber species.

As discussed in chapter 5, "Forest Clearance: Income, Technology, and Private Time Preference," integration to the market should produce non-linear effects on forest clearance. Integration with the outside economy through the sale of labor should reduce forest clearance because such integration will raise the value of people's time. Integration to the outside world through the sale of annual crops, however, should increase forest clearance. Since indigenous people sell annual crops and labor at the same time, integration into both markets should cause deforestation to resemble a non-linear form—such as a parabola.

The development of better credit and labor markets should reduce the need to swap goods and services with one's neighbor or kin because villagers will be able to borrow when faced with random, idiosyncratic shocks, such as illness or crop loss. Although the intuition has been around since the writings of the French sociologist Marcel Mauss (1990 [orig. 1927]), it has not been subject to a cross-cultural empirical test.

Economic development increases both the value of time (thereby reducing leisure) and income (thereby increasing leisure), leaving people with more or less leisure depending on the relative strength of the two forces.

The Ricardian model helps to generate several hypotheses about what might happen to indigenous people and their environment as they gain a firmer foothold in the market. The hypotheses include:

- more specialization in foraging because of the higher value of people's time and the change in relative prices between tradable and non-tradable food
- greater availability in the forest of forest goods that are replaced by commercial substitutes relative to the availability in the forest of forest goods exported from the village
- declining contribution of the forest to household income and consumption relative to income from wage labor and farming
- deeper knowledge of forest goods exported from the village and less knowledge of forest goods replaced by industrial substitutes
- non-linear effects of markets on the amount of forest cleared
- declining importance of reciprocity as new forms of insurance develop
- ambiguous effects of economic development on the quantity of leisure

Conclusion

Studies of why indigenous people take part in the market have gone through three stages. Some of the first researchers (Murphy and Steward 1956; Siskind 1973) alluded to the natural "process by which 'luxuries become necessities' and to the disruptive influence of trading relationships on native social organization" (Gross et al. 1979:1048). Starting in the mid-1970s, a second wave of researchers stressed the role of outside encroachers and government policies in pushing indigenous people to intensify production, degrade land, and turn to the market. A more recent group of scholars (e.g., Conklin and Graham 1995; Reed 1995; Smith and Tapuy 1995) emphasizes the conscious choice made by people in deciding to become part of market economies.

The three approaches are not mutually exclusive and complement each other. Circumscription may make indigenous people more receptive to market influences and make them more willing to sell goods and services. With circumscription, exposure to Western goods and outsiders becomes continuous and frequent rather than sporadic and brief (Wagley 1955). The lure of trade and industrial goods, always present, may gain prominence when people have less land and other natural resources.

Chapter 3

Research Design

Chapters 1 and 2 contained a discussion of the advantages and disadvantages of different methods used to study the effect of markets on welfare and on conservation among indigenous people. Some of the shortcomings included: failure to identify the direction of causality, reliance on bivariate analysis, and insufficient attention to definitions, to the role of omitted variables, and to functional form.

This chapter discusses how to rectify the shortcomings. The rationale for the choice of cultures, the methods used to collect information in each culture, the quality of the information collected, the implications of different measurement errors for the analysis, and the sampling strategy used are explained. The goal of the chapter is to clarify how the information was measured, collected, and analyzed.

Definitions, Causality, and Functional Form

This section explains the principles used to define variables, the techniques used to identify the direction of causality, and the statistical methods used to test for functional form.

Definitions

The main explanatory variable, integration to the market, was measured in four ways:

1. share of the harvest of selected annual crops sold,

2. earnings from wage labor,

3. value of credit received, and

4. total cash received from the sale of goods and services.

The collection of information was limited to annual crops (except with the Tawahka of Hondurus), because it is difficult in a cross-sectional study to capture the flow of perennial crops entering the household during an entire year with accuracy.

Information on integration to the market should have been collected for individuals because individuals buy, sell, and receive credit. A household could have some adults who sell a great deal and others who sell little. If people do not pool resources, the need to collect information about integration to the market at the level of the individual rather than at the level of the household becomes even more pressing. For this study, information from household heads (generally male heads) was collected and aggregated to the level of the household. This was done because a) it was assumed (erroneously perhaps) that households act as a unit, b) the procedure made it easier to collect information, and c) most of the researchers were men. Only among the Tawahka was socioeconomic information collected from individual women and men.

Since integration to the market overlaps with acculturation, knowledge of Spanish, reading, writing, and arithmetic were also measured. Tests were relied on, rather than self assessment measures, to reduce measurement error. Research subjects were also asked about the number of years of formal schooling they had completed.

Besides information on integration to the market and on acculturation, information on other explanatory variables, such as the demographic composition of the household, village attributes (e.g., proximity to old-growth forest), and household wealth, was also collected. Wealth was measured by combining the value of financial assets and liabilities and modern commercial assets (e.g., radios). For some dependent variables, such as deforestation and self-reported morbidity, we relied on people's own reports. For other dependent variables, such as body-mass index (kg/mt^2), time preference, and knowledge of plants and animals, people were measured, experiments were carried out, or tests were given. The same definition of variables was attempted across cultures, but some definitions were changed as time went on to make them more relevant to other cultures, to correct for earlier deficiencies, and to get more accurate information. For example, in the 1995 cross-sectional survey among the Tawahka, wealth was measured by assessing the financial value of pigs, chickens, cattle, and plastic buckets owned by the household, but later studies in Bolivia expanded the list to include the financial value of other, more traditional, physical assets (e.g., bows).

Causality

Researchers need to estimate the effect of integration into the market on different outcomes, but also need to control for the potential bias that may arise if the outcome affects integration to the market. To control for biases from potential reverse causality, the study used:

- lagged values for explanatory variables
- random variables over which the subject had no control
- instrumental variables

For instance, in the study of plant and animal knowledge (see chapter 11, "Of Trade and Cognition: On the Growth and Loss of Knowledge") information on many explanatory variables was collected a year before the collection of information on the dependent variable (knowledge). Illness in the household of a subject's neighbor was used when testing if markets weaken reciprocity, and the extent to which this exogenous shock affected the subject's savings in animals was estimated. Last, instrumental variables were identified and two-stage, ordinary least squares were used to make estimations. An instrumental variable is a variable that is highly correlated to an explanatory variable but contemporaneously uncorrelated to the disturbance term. For instance, in chapter 5, "Forest Clearance: Income, Technology, and Private Time Preference," a subject's own illness is used as an instrument for income to estimate the effect of income on deforestation.

Though adequate, none of the methods just listed guarantees eliminating biases from endogeneity. Lagged values of explanatory variables may not get rid of links between disturbance terms over time. Illness may serve as an adequate instrumental variable for income, but people may also select (albeit unconsciously) the optimal amount of health they wish to have. If mishaps strike an entire village, my neighbor's misfortunes and my misfortunes will be linked and both will affect my savings. For all these reasons, the methods used to correct for endogeneity should be read with caution.

Functional Form

Several of the following chapters test for non-linear relations between participation in the market (explanatory variable) and various outcomes. Chapter 5, "Forest Clearance: Income, Technology, and Private Time Preference," for example, tests whether forest clearance bears a parabolic (inverted U-shape) relation to economic development or cash income.

Rationale for the Choice of Cultures

The quantitative information presented in the following chapters comes from one culture in Central America (the Tawahka of Honduras) and from three cultures in the Bolivian lowlands (the Mojeño, Yuracaré, and Tsimane´). Information from the Chiquitano of Bolivia is used less frequently because the Chiquitano do not live in tropical rainforests. The total number of cultures studied was dictated by the money available for research and by the availability of students and colleagues to carry out the research.

The cultures were chosen because they had much variance in their degree of exposure to the market—the explanatory variable that matters most. One could have chosen other cultures, such as the Campa of Peru or the Yanomamö of Venezuela, because they also contain much variance in their degree of integration to the market. The cultures were identified and chosen because of the author's familiarity with the ethnography and history of the areas. Research in Central America should have taken place among the Sumu-Mayagna of Nicaragua, but rural unrest in Nicaragua forced the staff to move the research sites to Honduras and do the study among the Tawahka—the close linguistic and cultural neighbors of the Sumu-Mayagna.

In retrospect, the selection of cultures could have been better. The Tawahka may not have been as apt of a society for study as the Sumu-Mayagna. The Tawahka number only 900-1,000 people and do not display as much variance in degree of participation in the market as do the Sumu-May-agna, who number 10,000 people and who live in settlements spanning vastly different distances from towns and roads. Although the Bolivian cultures display much variance in exposure to the market, they lie close to each other (except for the Chiquitano) and have contact with each other, so they may not represent truly independent units of observation. Put more harshly, in the quantitative analysis of Part II one may be dealing with only three rather than five different cultures:

1. the Tawahka,
2. the Mojeño, Tsimane´, and the Yuracaré—representing only one unit of observation owing to their physical proximity, and
3. the Chiquitano.

With the benefit of hindsight, one ought to have chosen 3-4 non-contiguous cultures from Central and South America. One would have then had to weigh the benefits of the statistical analysis of working with a more scattered sample against the increased costs of doing such work.

Methods Used to Collect Information

Studies were designed in each culture to examine the effect of markets on the quality of life and on the use of the forest. Since the goal of each study differed, so did the methods used to collect information and the type of information collected (see table 3-1). Research was done with the approval of the indigenous government of each culture, often with research assistants provided by that same government. The purpose of the study was first explained to members of the indigenous government and, once approval was received, the purpose of the study was explained to villagers in community meetings and to individuals in one-on-one conversations. With the Tawahka, a contract was signed in which researchers pledged to deliver goods and services to the village and to the ethnic group in exchange for doing fieldwork.

Students collected information under the author's supervision and, in the case of the Tawahka, under the supervision of Nicholas Brokaw (plant ecologist) and David Wilkie (wildlife specialist). Graduate students had their own dissertation topics, which sometimes had a more historical and ethno-

TABLE 3-1 Summary of Methods Used to Collect Information

Ethnic Group	Sample and Methods					Dates of fieldwork
	Pop	% Pop	vil	hh	Methods	
Sumu-Mayagna	10,000	3	11	52	Participatory rural appraisal	6/92-8/92
Tawahka	900	13	5	15	Participatory rural appraisal	7/93-9/93
		88	5	101	Household surveys	6/95-8/95
		26	2	32	Time allocation	6/94-12/96
		26	2	32	Plant/animal census Weigh days	6/94-12/96
Yuracaré	3,339	12	11	62	Household surveys/ ethnography	6/95-8/95 and 8/97-8/98
Tsimane'	5,124	14	21	209	Household surveys/ ethnography	June-Aug., 1995 and 1996, and 8/97-8/98
Mojeño	19,759	4.3	17	132	Household surveys/ ethnography	6/95-8/95 and 8/97-8/98
Chiquitano	69,590	2.0	21	240	Household surveys/ ethnography	6/95-8/95 and 8/97-8/98

NOTES Pop is population, vil is village, and hh is household.

graphic slant. The author refrained from writing on topics when the interests of graduate students overlapped too closely with his own to allow students the chance of publishing first and of becoming first authors. The division of turf explains why this book draws gingerly on the material from the Tawahka.

Sumu-Mayagna (Nicaragua) and Tawahka (Honduras)

Nicholas Brokaw and David Wilkie helped design a study that was supposed to examine the effects of markets on the use of plants and animals by the Sumu-Mayagna of Nicaragua. The intent was to measure the effect of markets on foraging specialization, composition of household income, availability

of plants and animals in the forest, and the economic value of the forest measured by the non-timber goods removed (Godoy, Brokaw, and Wilkie 1995).

As part of a pilot study, the author spent about three months (June-
August, 1992) interviewing groups of people in 11 Sumu-Mayagna villages,
each with different levels of participation in the market. We found that as
incomes rose:

- the Sumu-Mayagna hunted fewer types of animals

- foraging played a smaller role in the household economy because people
  worked more in agriculture and outside of the farm

- the value of the forest, measured by the non-timber goods used, rose to the
  world median of $50/hectare (Godoy and Lubowski 1992)

A rise in income did not induce specialization in plant foraging. Education lowered dependence on the forest (Godoy 1994), a finding later con-
firmed in Honduras and Bolivia (Godoy and Contreras 2001).

Suitable research sites were found among the Sumu-Mayagna, but the
study could not be done because of rural unrest stemming from the after-
math of the civil war of the 1980s. The unrest prompted us to do a second
pilot study among the Tawahka of neighboring Honduras. The Tawahka
share cultural and linguistic affinities with the Sumu-Mayagna, but are
fewer and do not display as much variance in exposure to the market.

In 1993, Brokaw helped me carry out a pilot study among the Tawahka
to find new research sites. Two Tawahka villages, Krausirpe and Yapuwás,
were found along the Patuca River. The two sites shared the same culture,
history, and ecology, but differed in their level of income and in their distance to the nearest market town of Wampusirpe.

Research among the Tawahka of Yapuwás and Krausirpe unfolded over
2½ years (June 1994-December 1996) and drew on a sample of 32 house-
holds. Research was carried out by two graduate students in each village:
Josefien Demmer and Han Overman in Yapuwás, and Adoni Cubas and
Kendra McSweeney in Krausirpe. During the last eight months of fieldwork,
Glenda Cubas replaced McSweeney. The information on the Tawahka used
in this book stops in December of 1996—almost two years before Hurricane
Mitch hit the area.

Besides the core group of resident students, other students came for shorter
periods of fieldwork. During the summer of 1995, students Peter Kostishack
and Kathleen O'Neill from the United States (with Marques Martínez, a Miskito Indian working in a non-government organization in eastern Honduras)

did a quantitative study on the clearance of old-growth forest (Godoy et al. 1997). At the request of the Tawahka government, student Peter Cahn compiled oral histories from Tawahka elders and did a study of how the Tawahka's perceptions of the forest have changed since the beginning of the twentieth century (Cahn 1996, 1996a). In 1996, four more students (Daniel Colón, Susanne Lye, Adam Palermo, and Stanley Wei) did a census of forest plants and a survey of botanical and zoological knowledge (see chapter 11, "Of Trade and Cognition: On the Growth and Loss of Knowledge").

After selecting the two Tawahka villages, the group selected about 16 households in each village for detailed observations. During fieldwork, researchers monitored the flow of cash into households, did studies of time allocation, carried out census of plants and animals in the forest next to each community, and took inventories of physical and financial assets owned by the household. By following the same individuals over 30 months the group was able to create a short panel.

Weigh days were scheduled twice a month (on days chosen at random) to measure the flow of goods into the household. During weigh days, researchers identified, weighed, counted, and measured all goods entering the household from dawn to dusk, and asked where and how people had gotten the goods.

Information was collected on time allocation from spot observations and from focal follows. During spot observations (also known as scans), researchers wrote what every person over four years of age was doing at the moment they were first observed. Researchers did scans on non-overlapping time blocks of three hours (6am-9am, 9am-12pm, 12pm-3pm, 3pm-6pm). Scans were done during four days chosen at random each month. During focal follows, researchers followed a subject over 15 years of age from 6am until 6pm and noted the duration and the location of each activity of the subject.

Censuses of forest plants and animals were done to estimate the effects of extraction on the availability of resources in a direct way. To carry out the census of plants, 25 transects were set up in each village. Each transect measured 10 meters wide and two kilometers in length. The transects were set up perpendicular to the Patuca River, running up and down each side of the river. The total area of transects in each village was 50 hectares. With the help of Tawahka assistants, students walked the transects and recorded the abundance of plant species used by the Tawahka, including the diameter of the plant at breast height and the distance from the transect. Students also noted the vegetation type and the topography along the transect.

To count animals, we used three hunting trails radiating from each village. Each trail measured 3-4 kilometers in length. Students walked the trails at a constant pace at the same time of the day (5am-9am). During the walk, researchers and Tawahka guides noted the type and number of animals encountered, distance from the village, and vegetation cover at the location of sighting, time of sighting, and whether the animals were seen, heard, or smelled. Students also recorded signs of animal presence (e.g., footprints). Researchers walked trails a total of 71 times in the village of Yapuwás and 74 times in the village of Krausirpe. In the discussion of animal availability in chapter 6, "Game Consumption, Income, and Prices: Empirical Estimates and Implications for Conservation," only the information from direct observations of animals is included because it is more reliable than information from indirect signs (e.g. smell, footprints).

The survey on sources of cash income took place once a month. During the survey, students asked every adult in the household about the sources and the amount of income earned over the past month. To measure wealth, students carried out three inventories of selected physical and financial assets of the household, including the value of the house. During the surveys, students asked about the amount and the age of each asset. Through interviews with knowledgeable villagers, students got information on the useful life of assets and on the price or value of each new asset. The net, current worth of assets was estimated using straight-line depreciation. Since a few households also sold goods in small retail stores, income and wealth were estimated with and without the contribution of stores.

Early in the study, students carried out a demographic survey of all subjects in the sample. During the survey, they asked about standard demographic variables and human-capital attributes, such as formal schooling and knowledge of Indian languages. They monitored who left the sample so household size could be adjusted when making estimations with information from the panel.

Researchers entered information in solar-powered, portable computers while in the field. Computers allowed students to check information for accuracy and correct mistakes quickly. During the first six months of research, students from the two sites met periodically to enhance interobserver reliability in coding goods and behavior. The analysis does not include information gathered during the first six months of research because it was less reliable than information gathered after students had learned how to record information with accuracy.

The information from the Tawahka has the obvious advantage of being rich in detail, having been collected through direct measurements over 2½ years. The collection of detailed information from a few households in only two villages comes at the cost of a small sample size.

Another source of information for the Tawahka, and one used often in the following chapters, comes from a 1995 socioeconomic household survey of 101 Tawahka households in all five Tawahka villages. The sample covered most (88%) of the Tawahka population, estimated at about 900-1000 people (Caicedo 1993; Cruz and Benítez 1994). Because the Tawahka population is small—concentrated along five villages in one river—and because the movement of these people had been monitored for a year before doing the 1995 survey, it was clear who was absent at the time of the interview. Of the 14 households that were not interviewed, four lived outside the Tawahka territory, three did not want to take part in the study, and seven were probably missed because they were working outside of their village.

Mojeño and Yuracaré (Bolivia)

Two studies of Mojeño and Yuracaré groups were conducted, starting with a brief survey during June-August of 1995. During the second study, anthropology graduate student Tomás Huanca spent a year (August 1997-August 1998) doing an ethnography of the management of forests under fallow in Asunta—a Tsimane´ village in the headwaters of the river Sécure. During that time, he also did a household socioeconomic survey among the Mojeño, Yuracaré, and the Tsimane´ along the river Sécure (Huanca 1999).

During the pilot study of 1995, Jeffrey Franks (a macro economist), Mario Alvarado (a Bolivian undergraduate student working on his thesis), and the author did a survey among some of the three largest lowland groups in Bolivia (Mojeño, Yuracaré, and Tsimane´) in the plains of the department of Beni and along the river Sécure. The survey left out smaller lowland groups (e.g., Chácobo) because they did not display much variance in integration to the market.

With the help of the indigenous government of each ethnic group, a total of 23 villages were selected for preliminary research. Each government helped select three villages in their group along a continuum of integration to the market. For each village chosen, the indigenous government helped select a replicate with about the same degree of integration to the market.

Thus, for each ethnic group we had a total of about six villages straddling different amounts of participation in the market.

Tsimane´ (Bolivia)

After the 1995 pilot study with the Yuracaré, Mojeño, and Tsimane´, a second pilot study of just the Tsimane´ was conducted in 1996. The purpose of the second pilot study was to identify villages for a longer study and gain a better understanding of the effects of markets on health and the spread of vector-borne diseases, private time preference, forest clearance, and reciprocity. The second pilot study of the Tsimane´ was done with the cooperation of the Tsimane´ government (or Gran Consejo Tsimane´), and with Tsimane´ assistants assigned to us by the Tsimane´ government. Three undergraduate students—Vianca Aliaga and Julio Romero from Bolivia, and Joel DeCastro of the United States—helped with the study.

The Tsimane´ government was consulted to help identify villages with varying degrees of exposure to the market. With the aid of Tsimane´ assistants, the students interviewed and took anthropometric measures of 405 adults and 267 children in 209 households in 21 Tsimane´ settlements.

In May of 1999, a study of the Tsimane´ in two villages, San Antonio and Yaranda along the river Maniqui, began. The study ended in November of 2000, and focused on how markets affect the health and nutrition of children and adults, cultural consensus of botanical knowledge, economic vulnerability, and private time preference. That work was done by two graduate students in anthropology (Elizabeth Byron and Victoria Reyes-García), two undergraduate students from Bolivia (Lilian Apaza and Eddy Pérez), and by one Tsimane´ researcher (Alonzo Nate). Since the work was in progress when this book was written, the results are not included in the study.

Chiquitano (Bolivia)

Information for the Chiquitano comes from ethnographic fieldwork on local politics and a formal survey on the use of natural resources. Fieldwork took place from mid-1997 to mid-1998 and was done by Josh McDaniel—a graduate student in anthropology (McDaniel 2000).

Quality of Information

Information for many of the topics discussed in later chapters comes from surveys rather than from direct observations. One can ask about the size and the consequences of measurement errors for the analysis. To answer the question well would require comparing the information collected from surveys with the information from an objective yardstick. Since there is no such yardstick, the magnitude of measurement errors cannot be estimated. One example—in which what people said and what they did were compared—can be pointed to.

The example concerns the area of old-growth forest cleared by Tawahka households. Of the 101 households surveyed, 48 households had cut old-growth forest in 1994. Each household head was asked to estimate the area of old-growth forest they had cleared. Researchers with a tape and compass then measured each field. The results of the comparison showed that the Tawahka estimate field size with accuracy, and that the answers they gave in the surveys matched the measurement of the fields. The mean area of fields measured by the team of researchers was 0.98 hectares with a standard deviation of 0.72. The mean size of fields reported by the household head was 0.97 hectares with a standard deviation of 0.78. A two-sided, matched pair t-test of the equality of means produced a t value of 0.11, which falls within the 95 percent confidence interval (-0.14 to 0.15) for accepting the null hypothesis that there is no statistically significant difference between the area the Tawahka said they had cut and the area they had actually cut (Godoy et al. 1998a).

The finding does not mean that all the information collected through surveys is accurate. In the study of forest clearance, the Tawahka knew we were going to measure their fields after the interview so they may have tried to estimate with accuracy the size of their fields. They did not face the same incentives to report information with accuracy when asked about other topics because they knew we were not going to check their answers. Some topics are also harder to estimate with accuracy. Questions about income earned over the past month or year, for instance, requires that people recall information and make computations. Questions about income may therefore contain larger measurement errors than simpler questions about the size of fields. Measurement errors of income will be larger if people try to underestimate their income when answering survey questions.

Random measurement errors in dependent variables produce unbiased, but inefficient, estimates of parameters, reducing the statistical significance of results. On the other hand, poor but random measurement errors of explanatory variables bias the estimated coefficients toward zero. In either case, measurement errors will make it more likely to accept the null hypotheses that markets (or other explanatory variables) have no effect on an outcome when, in fact, they do. With random errors, the coefficients one estimates and the tests of statistical significance one carries out give a lower bound or a conservative reading of true magnitudes.

Sampling

For several reasons, random sampling was not used to select ethnic groups, villages, households, or people. First, it was more important to capture variance in integration to the market than to sample at random. Consequently, much effort was devoted to making sure villages (and people and households within villages) were selected that displayed variance in exposure to the market. As the information in table 3-1 suggests, we surveyed all five Tawahka villages, 11 Yuracaré villages, and between 17 and 21 Tsimané, Mojeño, and Chiquitano villages. For all cultures besides the Tawahka, at least two villages at low, middle, and at high levels of exposure to the market were captured.

Second, it is not clear how one samples households or people in ethnic groups where one lacks a baseline socioeconomic or demographic survey (Behrens 1990:307). The time and monetary costs of doing a baseline survey among an entire population before carrying out a formal socioeconomic survey are high. Third, over or under-representing an explanatory variable relative to its true value in the population has no effect on the estimated parameter for the explanatory variable.

Last, the household survey covered most (88%) of the Tawahka population. The estimated coefficients for the Tawahka can stand on their own (McCloskey and Ziliak 1996). Tests of statistical significance lose their normal meaning when one samples most of a population (Cowgill 1977:366-367). The tests then tell us more about the adequacy of the model than about statistical significance (Griliches 1986).

Although random sampling was not used to select households, random sampling was used to collect different types of information from households and to select subjects for different types of measurements. A random sample of people from our subjects in the village of Krausirpe was chosen, for example, to apply the test of knowledge of flora and fauna. To take anthropometric measures and apply the test of time preference to people who were not household heads, Tsimane' adults in the household who were not household heads were selected at random. Random sampling was used among the Tawahka to select observation periods for focal follows, scans, and weigh days. One of the two household heads among the Mojeño, Yuracaré, and the Chiquitano were chosen at random for the test of time preference.

The small sample size for each ethnic group and measurement errors should push one to pay more attention to the sign of the coefficients we estimate in Part II than to their level of statistical significance. The results of the comparative study should be looked at as suggestive rather than as definitive, as a guide for future research, or as a first step in a series of successive approximations to the truth.

One final point needs to be made about the consequences of the sampling strategy used. Because the information was collected in only two countries—Honduras and Bolivia—the effect of macroeconomic, sectoral, or international variables on the quality of life or on conservation cannot be estimated in a quantitative way. To do so one would have had to gather comparable information either from indigenous people in many nations or from the same people over a long period of time. This study did neither.

Conclusion

The comparative method used in this book has obvious disadvantages. Because a the author relied on a cross section of information, the dynamic processes cannot be estimated over time, nor can it control well for unseen fixed effects of people and households (except among the Tawahka). The historical and the ethnographic details so often described by ethnographers who work in one culture over many years cannot be captured. Because they were close to each other, some of the cultures under study may not represent truly independent units of observations—people from 2-3 cultures next to each other may all tell a similar story. Last, because the staff learned as they did the research, later surveys contain better questions and more accurate information than earlier surveys.

The methods used to collect and to analyze information, however, also have strengths. The comparative method allows for one to test hypotheses across cultures and draw tentative generalizations. Multivariate techniques allow researchers to control for the role of third variables, capture non-linearities, and identify the direction of causality in simple and direct ways. By including a total of about 65 villages, 700 households, and over 1,200 people, hypotheses can be rejected or accepted with more statistical confidence than is commonly the case in cultural anthropology. The weight of personal, household, village, and regional variables (but not national or international variables) can be estimated at the same time and thus link different levels of social organization in a quantitative way. Bearing in mind the strengths and weaknesses of the methods used, the next chapter turns to the ethnographic and historical record of each culture.

Chapter 4

Ethnographic Sketches

This chapter contains historical and ethnographic sketches of the Tawahka, Tsimane', Mojeño, Yuracaré, and Chiquitano. The sketches have been included for three reasons:

- to underscore each culture's long history of economic contact with the outside world
- to provide a context for understanding the statistical results of Part II
- to understand in a qualitative way why each culture displays variance in integration to the market

Many details have been left out of the vignettes because it was more important to highlight aspects of history and ethnography that bear in a direct way on later discussions. Descriptions of the habitat have been left out because information on the topic was not collected (except among the Tawahka) and because dummy variables for villages are often used in the statistical analysis to control for the effect of habitat. A village dummy variable is a binary variable that picks up all the unseen attributes of a locality, such as temperature, precipitation, altitude, and vegetation. To impose structure on the narrative and to make it leaner, ethnographic descriptions of dependent variables, such as health or private time preference, have been placed in the relevant chapters of Part II. To make easier the comparison, I have summarized some of the information presented in this chapter in tables 4-1 and 4-2.

TABLE 4-1 *The People: A Statistical Profile*

Variable	Tsimane' Obs	Mean	Sd	Mojeño Obs	Mean	Sd	Yuracaré Obs	Mean	Sd
Household Size:									
Adult men	237	0.94	0.81	132	1.31	0.57	62	1.14	0.43
Adult women	237	0.91	0.92	132	1.27	0.52	62	1.24	0.73
Children									
Boys	237	1.67	1.55	132	2.06	1.49	62	1.79	1.41
Girls	237	1.48	1.40	132	1.78	1.34	62	2.17	1.59
Total	237	3.16	2.32	132	3.84	2.13	62	3.96	2.14
Total	237	5.01	2.99	132	6.43	2.27	62	6.35	2.47
Forest (ha.) 1994:									
Old growth	237	0.63	0.74	131	0.41	0.49	62	0.40	0.58
Fallow	237	0.58	0.79	132	0.58	0.58	62	0.40	0.40
Market Links:									
% Rice	178	0.47	o.61	115	0.14	0.44	55	0.18	0.30
Days worked	209	54	89	132	37	74	62	29	54
% Credit	209	0.32	0.46	132	0.02	0.14	62	0	0
% Chemicals	194	0.06	0.25	132	0.32	0.47	60	0.24	0.43
Education:									
Male	207	1.91	2.71	132	2.85	2.24	59	1.93	1.77
Female	198	0.78	1.43	129	1.85	1.77	58	1.63	1.94
Spanish (%)	236	56.1	35.6	132	98.0	7.48	62	94.3	22.3
Distance (km):									
Mean		31.73		----------------- 123.08 ------------------					
Sd		18.31		----------------- 33.17 ------------------					

TABLE 4-2 *The People: A Statistical Profile (continued)*

Variable	Tawahka			Chiquitano		
	Obs	Mean	Sd	Obs	Mean	Sd
Household Size:						
Adult men	101	1.50	0.95	240	1.51	0.85
Adult women	101	1.60	0.92	240	1.33	0.67
Children						
Boys	101	2.39	1.50	240	1.75	1.52
Girls	101	2.24	1.44	240	1.56	1.33
Total	101	4.64	2.21	240	3.31	2.11
Total	101	7.75	3.34	240	6.16	2.59
Forest (ha.) 1994:						
Old growth	98	0.53	0.77	240	0.58	0.54
Fallow	98	0.95	0.95	240	0.69	0.47
Market Links:						
% Rice	91	0.07	0.12	147	0.03	0.10
Days worked	n/a	n/a	n/a	240	40	67
% Credit	101	0.16	0.37	240	0.10	0.30
% Chemicals	101	0.43	0.49	240	0.01	0.11
Education:						
Male	86	2.60	2.40	240	4.56	3.28
Female	15	1.40	1.90	240	2.72	2.45
Spanish (%)	85	54.7	44.0	240	97.5	12.6
Distance (km):						
Mean		37.67			65.49	
Sd		10.88			11.30	

NOTES Years for which information summarized: Tsimane' 1995; Mojeño, Yuracaré, and Chiquitano 1997; Tawahka 1994.
Under market links: numbers are per year; share of rice harvest sold, share of households receiving credit, share of households using chemicals for farming.
Education is maximum education of household heads.
Spanish is share of household heads who speak Spanish.
Distance from village to nearest market town estimated in straight line using global position system receiver. Reference points for distance: for Mojeño and for Yuracaré city of Trinidad; for Tsimane' town of San Borja; for Tawahka, town of Wampusirpe; for Chiquitano, town of Concepción. "n/a" means not available.

Tawahka

The first ethnographies of the Tawahka date to the early part of the twentieth century and were written by Eduard Conzemius (1932), a merchant and ethnographer who traveled widely through eastern Honduras and Nicaragua, and Francisco Landero (1935), a school teacher among the Tawahka. From the 1930s to the 1990s, the Tawahka received almost no attention from researchers. The flow of Nicaraguan refugees into Honduras during the Nicaraguan civil war (1980s) and concerns about Neotropical conservation probably explain the growth of interest in the Tawahka (Cruz and Benítez 1994, vol. 1, p. 157-158). In recent years, researchers have examined the Tawahka's use of plants (House 1997), land (Godoy et al. 1997), and other natural resources (Caicedo 1993; Herlihy 1997; Herlihy and Leake 1990, 1991, 1992; McSweeney 1999) and have tried to estimate the value of income imputed from foraging and farming (Padilla-Lobo 1995).

The Tawahka have lived in eastern Honduras and had links to the outside world for centuries (CIDCA 1982:24; Helms 1968:81; Davidson and Cruz 1988). During colonial days, they received English goods, salt, and beads from Misquito Indians in exchange for "canoes, paddles, gourds, calabashes, net hammocks, skins, and corn" (Conzemius 1932:40; Helms 1968:78, 1968a: 460, 1971:18).

During the nineteenth century, the Tawahka of the Patuca River exchanged dugout canoes for tools and for iron pots with outsiders (Young 1842:87 quoted in Cahn 1996:32). In the early twentieth century, the Tawahka and the Sumu-Mayagna—their cultural and linguistic neighbors in Nicaragua—tapped and sold rubber and game, collected and sold chicle and wild cacao, and traveled to work in logging camps and ocean-going boats (Conzemius 1932:40, 46-47, 64; Jenkins 1986; Landero 1935:6; Malkin 1956:165).

During the 1920s, the Tawahka started to get chemicals for farming from commercial banana plantations farther down the Patuca River (Cruz and Benítez 1994, vol. 2, p. 260). The flow of new farm technologies that started in the 1920s has continued intermittently to the present. During the 1980s, when refugees from the civil war in Nicaragua entered eastern Honduras, the Tawahka adopted improved varieties of rice and beans from international organizations and the army. During the 1980s, a local non-government organization started to distribute hybrid seedlings of cacao to people in the village of Krausirpe.

The arrival of Nicaraguan refugees in eastern Honduras in the 1980s affected the economy of the Tawahka and helped to create the political organization that today embraces all the Tawahka of Honduras. Most of the refugees from Nicaragua resettled just outside of the Tawahka territory, but the growth in the regional population produced by resettlement increased demand for crops and forest goods and put pressure on natural resources in the Tawahka territory. The influx of refugees also helped to heighten the political awareness of the Tawahka to events outside of their territory, and galvanized them to organize the Federación Indígena Tawahka de Honduras—the official political organization of all Tawahka (Cahn 1996). Since its creation in 1987, the Federación Indígena Tawahka de Honduras has:

- served as the official mouthpiece for the Tawahka
- tried to reduce encroachment by outsiders
- brought foreign aid and projects into the area
- gained official recognition for the territory of the Tawahka
- set up a program of bilingual education
- helped resettle the Tawahka who moved back into Tawahka territory from other areas

Although the territory of the Tawahka is relatively isolated (because it lacks roads linking it to western Honduras), it has faced encroachment from colonists and cattle ranchers moving east (Herlihy 1997). Small farmers and cattle ranchers from the department of Olancho have invaded the Tawahka territory to clear forest for farming or cattle grazing. It is still too early to tell whether the Federación Indígena Tawahka de Honduras will be able to stop future encroachment.

The Tawahka population has risen from 160 people to about 900-1,000 people during the twentieth century (Caicedo 1993). Much of the growth has taken place in the village of Krausirpe, where half of all Tawahka live. The rest of the Tawahka live in four smaller, poorer villages farther up the river. One such village is Yapuwás. In later chapters Yapuwás and Krausirpe are used as examples of a poor and rich village or as two villages with different levels of economic development. The socioeconomic differences between the two villages are discussed next to justify the selection of sites.

Krausirpe lies 30 kilometers from the market town of Wampusirpe in a straight line, and is the richest and most modern of the five Tawahka villages. Krausirpe contains offices for two non-government organizations, a government clinic, a primary and middle school, a church, and several retail stores—one of which villagers run as a cooperative. The more remote Yapuwás lies about 17 kilometers upriver from Krausirpe, and lacks stores, offices, or a clinic but has a primary school and a church. In mid-1995, Krausirpe had 58 households with 479 people and Yapuwás had 13 households with 91 people.

The information in table 4-3, drawn mostly from the 1995 household survey of all Tawahka, suggests that households in Krausirpe had 38.8 percent more imputed farm income than households in Yapuwás. The information collected from 32 households during 1995-1996 confirms the finding that households in Krausirpe had more income than households in Yapuwás. The information from the panel suggests that mean cash income of female and male household heads in Krausirpe were 2-3 times higher than in Yapuwás. The mean monthly cash income of female household heads in Yapuwás and Krausirpe, for example, were 79 and 204 lempiras (La). The results of an unmatched, two-tailed t test suggest that the difference was statistically significant at the 99% confidence level (t=5.39, p>|t|=0.001).

The two villages also differ in wealth. Information from the panel suggests that differences in mean quarterly wealth per person per household were large. The average person in Krausirpe had a mean stock of wealth of 1,536 lempiras, compared with only 238 lempiras for a person in Yapuwás (t=2.64; p>|t|=0.008). The information in table 4-3 suggests that households in Krausirpe had double the bean yields and 31% higher rice yields than households in Yapuwás. Households in Yapuwás used slightly more modern bean varieties (90% versus 75.87%) and sold a slightly higher share of their rice harvest (12.49% versus 8.27%). Households in Krausirpe, however, used a higher share of modern rice varieties (67.08% versus 37.55%) and sold a higher share of their bean harvest (11.17% versus 7.61) than households in Yapuwás. Households in both villages used about the same amount of chemical herbicides (one bottle).

The large difference in income and wealth between Krausirpe and Yapuwás and the strong presence of modern institutions in Krausirpe suggests that the choice of the two villages as proxies for rich and poor communities (or for two communities with different degrees of exposure to the market) was apt.

Tawahka subsistence centers on extensive swidden cultivation away from rivers and intensive farming by riverbanks. The Tawahka plant cacao and beans in the most fertile plots by riverbanks and put other perennials and

TABLE 4-3 *Yapuwás and Krausirpe Compared*

Variable	Krausirpe			Yapuwás		
	Obs	Mean	Sd	Obs	Mean	Sd
Farm Income						
Imputed	58	8137	6117	13	5862	4948
Monthly income/ person						
Wife	342	204	398	323	79	128*
Husband	351	1392	1876	275	415	766*
Quarterly wealth/person	135	1536	5362	120	238	258*
Yields						
Rice	56	8.40	5.43	12	6.38	14.40
Beans	49	5.84	4.14	10	2.85	1.68**
Modern farm technology						
Herbicides	56	0.99	2.18	13	1.19	1.14
Rice	55	67.08	37.88	12	37.55	43.30**
Beans	48	75.87	36.41	10	90	31.62
Crop sold (%)						
Rice	53	8.27	13.05	12	12.49	16.30
Beans	48	11.17	14.62	10	7.61	16.21

NOTES All values refer to 1994, except for quarterly or monthly information. Quarterly or monthly information refers to income or wealth/person, 1995-1996.
Income in *lempiras* (1 US$=9.40 *lempiras* in 1994).
Imputed farm income includes harvest of rice, beans, and cacao times village price.
Yields in *quintales/tareas*. 1 *quintal*=100 pounds. 1 *tarea*=0.25 hectares.
Assets refers to share of households owning modern assets (e.g., rifles).
Herbicides in bottles. Rice and beans under farm technology refers to share of seeds planted from modern plant varieties.
*significant at ≤5% in t test of comparison of means.

annuals—e.g. maize and rice—in upland plots cut from old growth or fallow rainforest (House 1997). The Tawahka use machetes, axes, and digging sticks to clear the forest, to plant, and to harvest. Although simple, their technological kit includes improved plant varieties and chemicals for farming. In 1994, 37 percent of Tawahka households used hybrid seedlings of cacao, 43 percent used chemical herbicides (see table 4-2), and 70-88 percent planted improved varieties of beans and rice (Godoy and Wong 2000). Besides farming, the Tawahka hunt, collect wild plants, and raise domesticated animals (House 1997; McSweeney 1999).

The Tawahka swap goods and services with each other in their villages, but need cash to buy food during the agricultural lean season, to buy school supplies and medicines, and pay for health services. They also need cash to buy food when mishaps strike (Godoy and Wong 2000).

In 1994, the average Tawahka household earned about 1,000 lempiras in cash income. Tawahka earned cash in several ways. Some worked outside their villages, panning for gold, making dugout canoes for sale, or helping on cattle ranches upriver (Padilla-Lobo 1995). A few worked for the Federación Indígena Tawahka de Honduras or for the central government as teachers. The sale of crops (other than cacao) was not an important source of cash. The average Tawahka household in 1994 only sold seven percent of its rice harvest (see table 4-2) and eight percent of its bean harvest (Godoy and Wong 2000). The sale of cacao had become the principal source of cash for villagers in Krausirpe only because Krausirpe has flood plains suitable for cacao production. Krausirpe had attracted the attention of a non-government organization that distributes hybrid seedlings of cacao and helps women produce and sell handicrafts.

Though brief, the description of the Tawahka highlights the difficulty of defining integration to the market through the use of only one criterion. Based on farm inputs, the Tawahka would appear to be closely linked to the outside economy through their intensive use of chemicals for farming—which goes far back in time. The survey done in 1995 showed that almost half (43%) of Tawahka households used chemicals. But if one defines integration to the market by the share of rice or of beans sold or by the amount of credit used, the Tawahka economy appears more autarkic. Only 16 percent of households received credit in 1994, and households sold only seven to eight percent of their rice or bean harvest. If one uses the sale of cacao to measure participation in the market, one would find a split population, with only Krausirpe linked to the market. In summary, the various forms through which integration to the market takes place requires: a) care defining terms, and b) the use of several definitions to ensure consistency robustness in the results of the analysis.

Tsimane'

Like the Tawahka, the Tsimane' first received attention from ethnographers in the early part of the twentieth century (Nordenskiold 1924; Métraux 1948), but they did not become the subject of interest to social (Chicchón 1992; Daillant 1994; Ellis 1996; Huanca 1999; Piland 1991; Riester 1993) or natural scientists (Gullison 1995; Gullison et al. 1996; Howard, Rice, and Gullison 1996; Rice, Gullison, and Reid 1997) until the 1990s.

One of the largest indigenous groups of lowland Bolivia, the Tsimane' number 5,124 people and live on the plains and in the rainforests of the department of Beni (Censo Indígena 1994-1995). The Tsimane' population extends over several municipalities and four political jurisdictions—Pilón Lajas Reserve, Territorio Uno, Territorio Multiétnico, and Parque Nacional Isiboro Sécure. Like other Bolivian lowland groups, the Tsimane' lack a central government representing the interests of the entire ethnic group. The Tsimane' government (or Gran Consejo Tsimane') primarily represents the Tsimane' of Territorio Uno, but also tries to represent the interests of Tsimane' from other areas.

The Tsimane' avoided permanent contact with missionaries, traders, and outsiders until the late nineteenth century (Castillo 1988; Ellis 1996:16-17; Nordenskiold 1924; Pérez-Diez 1984). For reasons that are still unclear (that perhaps have to do with pressure from encroachers), the twentieth-century Tsimane' started to pan gold, extract quinine, tap rubber, sell rice and pelts, and work as unskilled laborers on cattle ranches and in logging camps (Piland 1991). In 1953, Redemptorist missionaries set up a mission in the upper reaches of the Maniqui river (Ellis 1996:18). Protestant missionaries also entered the region in the 1950s, setting up schools and a clinic, and training many of today's top Tsimane' political leaders. Pressure from encroachers increased in the early 1970s. Traders from the towns of Yucumo and San Borja started to enter the Tsimane' territory in the 1970s, giving credit and alcohol in exchange for rice and forest goods. Highland colonists also started to invade the area in the 1970s (Riester 1993). The role of Protestant missionaries as brokers for the Tsimane' has declined—overshadowed by the increasing autonomy of the Gran Consejo Tsimane', which the missionaries helped to create in 1989.

Contact with the market and acculturation have not caused social disorganization, nor have they caused the Tsimane' to take part in messianic movements, as they have among the neighboring Mojeño and Yuracaré

(Lehm 1991; Riester 1976). Instead, contact seems to have induced Tsimane´ to plant only a few crops, deplete game (Ellis 1999), reduce the length of fallow, mine the soil (Piland 1991), and sell valuable goods from the forest (Añez 1992; Gullison 1995; Gullison et al. 1996; Howard, Rice, and Gullison 1996; Rice, Gullison, and Reid 1997; Rioja 1992).

The Tsimane´ take part in markets in different ways and degrees. Remote villages contain monolingual speakers of Tsimane´ who forage and practice swidden farming. Their contact with outsiders is limited to bartering rice or thatch palms for salt, metal tools, and alcohol. Villages closer to towns have ham radios and primary schools with bilingual teachers. These villagers find it easier to sell rice, buy commercial goods, and work in nearby cattle ranches and logging camps.

The information in table 4-1 suggests that the Tsimane´ of Territorio Uno are well linked to the market in many ways. Unlike the Tawahka who only sell six to seven percent of their rice and bean harvest, the Tsimane´ sell almost half (47%) of their rice harvest. Tsimane´ household heads also work in wage labor 54 days each year—more than any of the other groups studied. Thirty-two percent received credit or advances on salary from loggers, ranchers, or town merchants. Unlike the Tawahka, few Tsimane´ use chemicals for farming—only six percent of the households interviewed had used chemicals.

Mojeño and Yuracaré

The Yuracaré and the Mojeño share many socioeconomic similarities because they live next to each other—often in the same village—but they differ in history and culture.

The Yuracaré and the Mojeño villages studied in 1995 lie along several parallel rivers, which drain the Eva Eva Mosetene basin before discharging their waters into the Mamoré River. Except for a few villages that abut or include savanna, most of the villages studied were in old-growth rainforest. For the statistical analysis of Part II, only information from Mojeño and Yuracaré households of the river Sécure (rather than from the plains) is used because those households live in areas with old-growth rainforest. This makes it easier to control for ecology when comparing them to the Tsimane´ and the Tawahka.

The Mojeño and the Yuracaré forage, but they rely on swidden farming along riverbanks for most of their subsistence, and use digging sticks and metal tools to prepare their land (Alvarado 1996). Their cash crops include rice, manioc, and fruits. The Yuracaré of the river Sécure sell a slightly higher share of their rice harvest (18%) than do the Mojeño (14%) (see table 4-1). Mojeño and Yuracaré villagers along the plains also sell firewood, logs, and thatch palm. Since the floods of 1992-1993, ranchers, loggers, missionaries, and non-government organizations have brought chemicals, chain saws, modern rice seeders, and new varieties of rice, maize, and pasture grasses (Lehm 1994). Mojeño and Yuracaré innovators are experimenting with the cultivation of beans, wheat, vegetables, and soybeans. Twenty four percent of the Yuracaré and 32 percent of the Mojeño households surveyed in 1997 had used chemicals for farming. Only two percent of Mojeño households (and no Yuracaré household) received credit in 1997 (see table 4-1).

Besides earning cash by selling crops, the Mojeño and Yuracaré also earn cash by working in logging camps, on cattle ranches, and in towns. This type of wage labor occurs during the dry season—from May until November. During 1997, typical Yuracaré and Mojeño household heads worked 29 and 37 days in wage labor (see table 4-1).

Paralleling those similarities are differences in demography, history, economics, politics, and culture. There are 19,759 Mojeño and 3,339 Yuracaré (Molina 1994). The Mojeños have had centuries of contact with Catholic missions and cattle ranchers (EPRM 1989:82-83; Jones 1980, 1991, 1995; Roper 1999). A longer history of symbiotic (though at times hostile) contact helps to explain why Mojeño villages in the plains have more people, stores, government offices, and more out-migrants than Yuracaré villages (Godoy, Franks, and Alvarado 1998). Most Mojeño outside of the river Sécure speak Spanish and benefit from the agricultural extension, adult education, and credit offered by non-government organizations and from the proximity to the town of San Ignacio de Moxos. The information in table 4-1 suggests that adult Mojeño men have more education (2.8) than adult Yuracaré men (1.9), but the difference is less prominent among the women of the two groups (1.8 versus 1.6).

In contrast to the history of the Yuracaré, the history of the Mojeños is peppered with messianic movements going back to the nineteenth century (Lehm 1991; Riester 1976). Three of the more isolated Mojeño villages studied in 1995 were spawned out of recent messianic movement. During these episodes, villagers give up all trappings of the modern world—including clothing, tools, and medicines—and move farther into the forest in search of a better way of life in a promised land stocked with cattle but without outsiders (Lehm 1991).

In contrast to the Mojeño, the Yuracaré have had less contact with missionaries or cattle ranchers—in part, perhaps, because they have been more skillful at avoiding them (Miller 1917; Paz Patiño 1991; Ribera 1983). When Mather (1922) canoed through the river Sécure in the first quarter of the twentieth century, he found no Yuracaré (or any other) village along the river. This suggests that the presence of Yuracaré in this remote area may reflect attempts to escape from the encroachment of highlanders and cattle ranchers (Ribera 1983).

Like the Mojeño, the Yuracaré of the river Sécure are bilingual and their more accessible villages benefit from agricultural extension, adult education, and credit offered by non-government organizations. Unlike the territory of the Mojeño in the plains, the territories of the Yuracaré and the Mojeño in the river Sécure are intentionally off limits to the rest of the world. The authorities of the Parque Nacional Isiboro Sécure have vetoed demands by Yuracaré and Mojeño villagers along the river Sécure to build feeder roads to logging camps or to the city of Cochabamba for fear of luring coca cultivators (Thiele, Johnson, and Wadsworth 1995). The authorities of the Parque Nacional Isiboro Sécure have also banned private traders from entering the river Sécure because they say traders overcharge for the goods they sell and pay too little for the goods they buy.

Chiquitano

Like some of the groups discussed earlier, the Chiquitano fought against the domination of outsiders and retreated to inaccessible regions to avoid encroachers (Krekeler 1995; McDaniel 2000). At present, most Chiquitano live in the dry, tropical, semi-deciduous forest of the department of Santa Cruz. A few Chiquitano live in the department of Beni. There are 69,590 Chiquitano, making them the largest indigenous group in the Bolivian lowlands (Censo Indígena 1994-1995; Rozo 1999; Schwarz 1993:96).

Once resettled in Jesuit missions, the Chiquitano started to produced wax and cotton cloth for the silver mines of Potosi (Riester and Suaznabar 1990:6). During colonial times, they raided Spanish settlements to get iron tools and soon grew accustomed to their use (Krekeler 1995:216). After the expulsion of Jesuits in the late eighteenth century, outsiders started to move into the Chiquitano territory and the Chiquitano started to work for them as unskilled laborers (Schwarz 1993:19). By the late nineteenth century, the Chiquitano were being recruited to tap rubber outside of their territory (Riester and Suaznabar 1990:7; McDaniel 2000). In the mid-1950s, with the decline of rubber, the Chiquitano were recruited to build the railway from Santa Cruz to Corumbá. Wealth from the rubber boom helped to finance many of the sugar plantations set up in the Chiquitano territory (Schwarz 1993:21). Starting in the 1970s, logging firms and private entrepreneurs moved in to extract valuable timber species in the Chiquitano territory. Menonnite missions and cattle ranches were also set up or expanded, which reduced the amount of land available to the Chiquitano for subsistence.

Chiquitano subsistence centers on swidden cultivation in old-growth forest, fishing, and (to a lesser extent), hunting and plant collection (McDaniel 2000). Staple crops include rice, manioc, peanuts, maize, sweet potatoes, plantains, and fruits. Wild honey is one of the most important non-timber forest products collected. Fishing becomes the predominant activity in the dry season, and accounts for a quarter of the yearly value of household consumption (Riester and Suaznabar 1990:39). Fishing is less important in communities next to cattle ranches because of restrictions on fishing imposed by outside landowners (Riester and Suaznabar 1990:36).

The Chiquitano earn cash through wage labor and the sale of farm goods. Nearby cattle ranches, commercial farms, private logging firms, and a logging cooperative run by the Chiquitano provide the Chiquitano with employment during the agricultural slack season. The 1998 survey shows that Chiquitano household heads worked 40 days a year in wage labor— one of the highest figures among the five cultures (see tables 4-1 and 4-2). Chiquitano also earn cash by selling peanuts and wild honey. A quarter of the households sold peanuts. Households sold an average of 72 percent of their peanut harvest. Households only sold three percent of their rice harvest. The sale of maize and rice may be displacing the sale of peanuts as a source of cash (Rozo 1999). Ten percent of Chiquitano households had access to credit.

Chiquitanos use cash to buy clothing, tools, meat, sugar, lard, kerosene, and school supplies. Only one percent of the sample surveyed in 1998 had used cash to buy chemicals for farming in 1997 (see table 4-2). Communities closest to towns and those that have suffered most from the encroachment seem to buy and to sell more (Schwarz 1993:88-89). Even remote communities have been drawn to the market because a growing number of peripatetic traders from the highlands have started to comb through the Chiquitano territory with regularity.

Similarities and Differences

Similarities

All the cultures rely primarily on swidden cultivation and, secondarily, on foraging. They inhabit tropical rainforest, although some live in savannas or deciduous forest. The information in table 4-1 suggests that households clear about half a hectare of old-growth forest a year, but 37 percent of households did not cut old-growth forest. Except for the Tawahka, households clear about the same area of old and of secondary-growth forest. The Tawahka clear twice as much fallow as old-growth forest—perhaps because of greater population pressure.

The cultures also resemble each other in settlement pattern. People in all cultures live in nucleated settlements, ranging from a low of eight households to a high of 90 households per settlement. Each culture displays variance in village density.

All the groups have had centuries of contact with outsiders, although contact seems to have been strongest and more continuous among the Mojeño and the Chiquitano. Many of the cultures have moved to inaccessible regions to avoid or minimize contact with outsiders. Except for the Mojeños and the Yuracaré of the river Sécure, all the other groups face direct threats from encroachers.

In recent years, the groups have formed new political organizations to help defend their territorial rights. The Federación Indígena Tawahka de Honduras, the Gran Consejo Tsimane´, and the umbrella organizations representing the Mojeño, Yuracaré, Tsimane´, and Chiquitano (Central de Pueblos Indígenas del Beni and Confederación Indígena del Oriente, Chaco y Amazonia de Bolivia) were all created in the late 1980s. The Federación Indígena Tawahka de Honduras has worked with environmental organizations to stop the building of a hydroelectric dam on the Patuca River, and has successfully lobbied the central government to hire bilingual teachers in the Tawahka territory. In the early 1990s, lowland groups in Bolivia marched to the nation's capital (La Paz) to demand communal titles to their lands. In response to such pressure, the government of Bolivia has started to grant indigenous lowland people title to their land, has put in place a bilingual system of education, and has given municipalities greater responsibility in administration and finance.

People in all the cultures depend on the market to various degrees, and need cash to buy school supplies or necessities such as salt and metal tools. Cash becomes indispensable when misfortunes strike. To satisfy their need for cash, people in all cultures sell goods such as beans, peanuts, rice, thatch palm, logs, game, wild honey, and firewood. During the agricultural slack season, they work as unskilled laborers on cattle ranches, in logging camps, in the farms of smallholders, and in nearby towns. Private merchants comb through indigenous territories (except along the river Sécure), selling commercial goods, medicines, and farm inputs, buying crops and non-timber forest goods, and giving credit to people they know.

Differences

The groups vary in population size—from a low of 900-1,000 (Tawahka) to a high of 69,590 (Chiquitano). They also vary in household composition and human capital. Total household size increases as one moves from the Tsimane´ (5.01), to the other groups in Bolivia (6.2), and on to the Tawahka (7.7). The greater household size of the Tawahka reflects a greater number of children. The people in the sample have little formal schooling—averaging only about two years. The Chiquitano have twice as much formal schooling than the rest. The groups fall into two clusters in their knowledge of Spanish. Among the Tsimane´ and Tawahka, only about 55 percent of household heads speak Spanish with fluency. Among the other groups, almost all household heads speak Spanish fluently.

The groups also differ in political organization. Through the Federación Indígena Tawahka de Honduras, the Tawahka have been able to present the outside world with a unified political front. The groups in Bolivia, being larger and more dispersed, have been fractured by internal differences and have had to rely on umbrella political organizations that embrace many ethnic groups.

The groups vary in distance to the nearest market (table 4-1). The Yuracaré and the Mojeño of the river Sécure are the most isolated—their villages lie an average of 123 kilometers from the nearest town.

Although all groups face the threat of encroachment, principally from smallholders, the pressure seems most intense in the territory of the Tawahka—where conflicts have erupted with cattle ranchers, smallholders, and people displaced by the Nicaraguan civil war (Herlihy and Leake 1991, 1992; Herlihy 1997). In Bolivia, the alliance between cattle ranchers and colonists from the highlands and indigenous peoples goes back many years and does not seem to be marked by the kind of conflict seen in Central America. The land of indigenous people in the river Sécure still lies outside the direct threat of coca cultivators and other encroachers. The same cannot be said of the land of the Tsimane' or the Chiquitano. Logging firms have perforated the territory of the Tsimane'—taking out logs regularly and with impunity, and often with the help of the Gran Consejo Tsimane' and ordinary villagers. Commercial logging has not produced conflict with outsiders, although it has pitted villages against the Gran Consejo Tsimane' as each tries to gain a bigger share of the economic rents from logging.

The cultures under study have experienced different shocks in recent years. The Tawahka did not experience the civil war in a direct way, but saw the consequences of war in the wave of refugees that crossed over to Honduras. The floods of 1992-1993 affected the Tsimane', Yuracaré, and Mojeño. The territory of the Tsimane', Yuracaré, and Mojeño was also a pit stop for drug traffickers during the 1980s. Drug trafficking has declined in recent years owing to a stronger program of drug interdiction. Although the production and the distribution of coca and cocaine have affected most aspects of the Bolivian economy (DeFranco and Godoy 1992; Gibson and Godoy 1993), it does not presently affect the household economy of the groups in this study in a direct way.

Last, the groups differ in how they link to the market. The information in table 4-1 suggests that the Yuracaré and the Mojeño are well linked to the market through the sale of labor, but not through the sale of crops or the use of chemicals or credit. The Tsimane' are poorly linked to the outside economy through the market for farm inputs, but seem well linked through the markets for labor and for rice. The information for the Tawahka would suggest an economy that is poorly linked through the output or credit market, but well linked through the market for farming chemicals. The Chiquitano are tied to the outside economy chiefly through the sale of labor.

Conclusion

Despite differences, the cultures under study show a long history of economic contact with the outside world, many attempts to fight or avoid the encroachment of outsiders, and different ways of linking with the outside economy. Ties to the outside economy include the sale of labor, forest products, and crops, and the use of credit and modern farm inputs. The empirical analysis of how markets have affected the use of natural resources and the quality of life comes next.

Part II

THE FINDINGS

The chapters of Part II explore the relation between markets and 1) the use of forest goods and demography (chapters 5-7), 2) welfare (chapters 8-10), and 3) knowledge of forest plants, game, and private time preference (chapters 11-12).

The chapters on the use of forest goods and demography explore the effect of markets on the clearance of old-growth forest, the consumption of game, and demography. Chapter 5 analyzes how income, farm yields, and private time preference affect the area of old-growth forest cut by households. It tests whether economic development causes forest clearance to resemble an inverted, U-shaped (or Kuznets) curve, whether higher crop yields reduce the amount of forest cleared, and whether private time preference has the effects on forest clearance predicted by economic theory. Chapter 6 estimates the effects of income, the price of game meat, and the price of meat from domesticated animals on the consumption of wildlife. It then estimates the effect of economic development on the availability of game, by comparing the likelihood of game sightings in the forests of a poor and a more prosperous Tawahka village. Chapter 6 contains one of the first estimates of the income, own-price, and cross-price elasticities of game consumption. Chapter 7 estimates how taking part in the market may affect the role of demography in production. The conditions under which Chayanov's theory of a household developmental cycle holds up are also identified.

Chapters 8-10 explore how markets affect welfare. Chapter 8 tests Sahlins's venerable hypothesis about the decline of leisure as economies modernize. The hypothesis is tested using multivariate techniques on a short panel of information on spot observations of the Tawahka. A panel consists of repeated observations of the same people, households, and localities over time. Chapter 9 examines how markets and acculturation affect self-perceived and objective health and Chapter 10 tests whether Mauss' idea that reciprocity weakens with modernization holds up to cross-cultural evidence.

The last part of Part II (chapters 11-12) analyzes how markets affect people's knowledge of forest plants and game (chapter 11) and explores the causes and consequences of private time preference (chapter 12). Private time preference or private discount rate refers to a person's willingness to delay gratification for future rewards. The sentiment affects how much people save, consume, and invest, and shapes how an entire economy operates.

The chapters of Part II form the core of the empirical analysis and set the stage for the conclusions of Part III.

Chapter 5

Forest Clearance: Income, Technology, and Private Time Preference

The debate about how markets affect conservation has divided researchers into three camps. Anthropologists since the days of Daniel Gross (et al. 1979) and Shelton Davis (1977) have said that markets and state policies hurt conservation (Bodley 1988; Painter and Durham 1995; Sponsel, Headland, and Bailey 1996). A second group of scholars has said that markets enhance conservation, provided people enjoy secure rights of property to land (Hyde, Amacher, and Magrath 1995; World Bank 1992). A third group of scholars has said that economic development produces ambiguous effects on conservation (Bawa and Dayanandan 1997; Cropper and Griffiths 1994). Although social and natural scientists have been studying the causes of tropical deforestation since the early 1970s (Denevan 1973; Gómez-Pompa, Vazques-Yáñez, and Guevara 1972), they have been unable to reconcile divergent views on the effects of economic development on the loss of rainforest.

This chapter tries to accomplish three goals. First, a model (developed with D. Wilkie and J. Franks) detailing how markets affect the loss of old-growth rainforest is presented. Five hypotheses from the model and empirical evidence from five cultures to test the hypotheses are also explored. The conditions under which markets hurt or help conservation are explained to bridge opposing views in the debate. Second, the chapter tests the hypothesis that higher crop yields reduce deforestation. Last, it examines how private time preference affects forest clearance. The terms private time preference, private discount rate, and patience are used interchangeably. The concepts are equated with a person's willingness to substitute consumption over time or delay gratification. Although private time preference is discussed in chapter 12, one of its ecological consequences is examined in this chapter to assess the weight of a psychological determinant of behavior.

Rationale for the Choice of Indigenous People and of Old-Growth Forest to Study Deforestation

Choosing indigenous people to study Neotropical deforestation does not mean that they are the main culprits. This study focused on indigenous people for three reasons. First, indigenous people have received less attention than loggers, cattle ranchers, commercial farmers, or smallholders in studies of deforestation. It still isn't known with certainty who is responsible for most of the rainforest loss in the tropical lowlands of Latin America. Researchers and policy-makers blame smallholders, ranchers, commercial farmers, and loggers (Hecht 1998; Sierra and Stallings 1998).

Second, although forest clearance by indigenous people is not an environmental threat today, it will become one in the future. Growth in population and income inside indigenous territories will put more pressure on natural resources (Greenbaum 1989; Vidal 1989; Picchi 1991). Logging or forest clearing by indigenous people may not result in large areas of forest loss at any one time. As discussed in the last chapter, the average household in the sample cut only about half a hectare of old-growth or secondary-growth rainforest each year. Over time and at current rates of population growth cutting will increase forest fragmentation and the loss of forest-dependent species. Even low rates of cutting could threaten the survival of small, indigenous reserves. As indigenous people gain greater sovereignty over their reserves, they will need to take steps to lower internal pressure on their forests. Curbing commercial farmers and cattle ranchers from cutting forest was yesterday's problem; curbing indigenous people from mismanaging their natural resources will likely be at the top of tomorrow's policy agenda.

Third, indigenous groups with large variances in integration to the market provide an ideal laboratory for testing hypotheses about the role of markets on deforestation. The hypotheses cannot be tested as well with other users of the forest because other users are already well linked to the market.

This study concentrates on the loss of old growth rainforest rather than secondary-growth forest because old-growth rainforest contains more biological diversity and because it lies at the center of the policy debate among conservationists. Secondary-growth forest can also be rich in species and contribute to carbon sequestration (Denevan 1992; Finegan 1996; Frumhoff 1995; Lawrence, Leighton, and Peart 1995; Saldarriaga et al. 1985; Silver, Brown, and Lugo 1996). As shown below, the results of the analysis presented in this chapter apply to both types of forest.

The Model

This section presents a model of deforestation to: 1) avoid producing *ad hoc* statistical correlations without theoretical foundations, and 2) draw attention to a few explanatory variables that matter most. The process by which a household becomes integrated into the larger economy is modeled in the same way trade theorists model the integration of national products and factor markets into the international economy. A qualitative explanation of the model follows.

Income From Annual Crops

As rural households become part of market economies, the demand for their annual crops increases from an increase in income and population (Netting 1982, 1993). Traditional households clear small plots of forest to farm because they have simple technology and only need to satisfy local demand. As markets envelope autarkic households and households adopt new technologies to clear the forest, external demand for crops grows, causing an increase in exports of annual crops. The increase in external demand for annual crops should cause forest clearance to increase (Behrens 1992a; Ehui, Hertel, and Preckel 1990).

Income from Wage Labor

As rural households become part of market economies, demand for their labor also increases. Remote villages export few workers to the outside world. People from those villages earn only a small share of their income from jobs outside the farm (Werner 1979; Stier 1982, 1983). When villages tighten their links to the market, the demand for workers to extract forest goods or work on plantations or ranches grows, increasing the value of rural people's time and lowering their incentives to work on the farm, to practice conservation, or to clear forests (Behrens 1992b; Homma 1992; Burkhalter and Murphy 1989).

Total Income

The offsetting effect of taking part in the market for annual crops and for labor should, at the same time, cause forest clearance to bear a non-linear relation with total cash income—defined as the sum of farm and labor income. Non-linear relations take many forms, including a U-shaped, or an inverted U-shaped, parabola. Economists have said that economic development produces a Kuznets curve of environmental degradation (Arrow et al. 1995; Grossman and Krueger 1995; Stern, Common, and Barbier 1996). A Kuznets curve resembles an inverted U. As national economies increase their income and their links to the world market, deforestation increases first and falls after it reaches a threshold of income. The model examined in this chapter suggests that economic development produces a non-linear effect on forest clearance, which may or may not resemble a Kuznets curve.

Crop Yields and Private Time Preference

The model just sketched helps to explain the effect of markets on forest clearance, but it does not explain the role of crop yields and private time preference on deforestation. These two variables are discussed for reasons of public policy and anthropological theory. If increased crop yields reduce forest clearance, then policy-makers and indigenous people will have one more policy lever for improving conservation. The role of private time preference in deforestation matters because it allows one to estimate the weight of a psychological determinant of behavior while controlling for confounding covariates.

The effect of crop yields on forest clearance will be ambiguous because it depends on whether people consume or sell the crop. An increase in the yield of a non-traded crop used chiefly for subsistence in the village will increase the supply of the crop, reduce the value or the cost of production, and reduce the area of forest cleared. Improvements in the yields of a crop exported from the village, however, should increase producer surplus in unambiguous ways—inducing producers to expand their fields.

Private time preference could affect forest clearance in two opposite ways, depending on how one views forest clearance. If one views forest clearance as a form of private investment, through which people build inheritance for their heirs, accumulate wealth, or lay claims to land (Angelsen 1999), then a low rate of private time preference should cause forest clearance to increase. Among the groups under study, forest clearance resembles private investment because some households clear forest to leave as inheritance for their children. One could also view forest clearance as a form of immediate consumption rather than investment. As people clear forest they can sell logs and use the remaining branches as firewood. Some annual crops only take three to four months before people can harvest them. If people convert forests to pasture, herbivore-carrying capacity will increase and so will the availability and revenues from selling livestock (Robinson and Bennett 1999).

Since forest clearance contains a bit of both investment and consumption—because people can sell or use timber and firewood from the plot immediately (consumption) before burning the forest for agriculture and staking claims to the land as it reverts back to fallow (investment)—theory suggests that the effect of private time preference on deforestation ought to be ambiguous, and ought to depend on the relative strength of the investment and the consumption effects.

The idea that private discount rates could enhance or worsen deforestation fits with the current work of Bohn and Deacon. Drawing on the theoretical work of Y. Hossein Farzin (1984), on quantitative information over time from a cross-section of nations and qualitative historical information from past civilizations, Bohn and Deacon (Bohn and Deacon 2000; Deacon 1994, 1999) present evidence to suggest that periods of political turbulence correlate with greater depletion of some types of natural resources and with reduced depletion of other types. They find that high discount rates accelerate the depletion of natural resources when investment costs are low. If the extraction of a natural resource requires high initial investments, then political uncertainty tends to curb extraction.

Bohn and Deacon go on to show that political uncertainty curbs the extraction of oil, gas, and hard minerals because the resources require large outlays in exploration and development before production can begin. They also show that political uncertainty hastens the clearance of forests in most developing nations because people and firms there tend to clear forest without making large capital investments. Much deforestation in developing nations reflects the actions of poor farmers working with simple technologies. In forestry, unlike mining or oil exploration, high discount rates increase depletion because investment costs are less important.

Hypotheses

The model and previous discussion lead to the following hypotheses:

Hypothesis 1: Integration through the labor market will lower the clearance of old-growth forest. The amount of annual income earned from wage labor, or the number of days a year people spend working in non-farm jobs, will be associated with a smaller area of old-growth forest cleared.

Hypothesis 2: Integration through the market for annual crops will increase the clearance of old-growth forest. The amount of annual crops produced will bear a positive correlation with the area of old-growth forest cleared.

Hypothesis 3: Integration through both markets—or total income from the production of annual crops and wage labor—will bear a non-linear relation to the clearance of old-growth forest.

Hypothesis 4: Crop yields will have opposite effects on forest clearance, depending on whether crops are non-traded and used for household consumption (less clearance) or tradable and used for export (more clearance).

Hypothesis 5: Private time preference will have ambiguous effects on forest clearance because forest clearance could be a form of consumption or investment.

Previous Studies

Before turning to the empirical analysis, this chapter reviews quantitative studies on the relation between forest clearance and economic development or income, crop yields, and private time preference.

Economic Development or Income

Previous household studies seem to confirm the first two hypotheses. Results of household studies in developing countries show that higher wages reduce deforestation and that higher crop prices increase it (López 1993a, 1993b). A case study from Peru suggests that non-farm work also reduces forest clearance (Bedoya 1991, 1995).

Studies on the link between deforestation and total income have produced unclear results—as predicted by the third hypothesis. Case studies in Puerto Rico, India, Kenya, and other regions show a rise and decline in forest clearance as incomes grow (Foster, Rosenzweig, and Behrman 1998; Hyde, Amacher, and Magrath 1995; Lugo 1992; Patel, Pinckney, and Jaeger 1995; Rudel 1998; Ruitenbeek 1988, 1989; Singh 1999). International comparisons have also uncovered evidence for other forms of non-linearities (Allen and Barnes 1985; Rudel 1989; Rudel and Horowitz 1993; World Bank 1992). Cropper and Griffiths (1994) did a multivariate analysis of the causes of deforestation in a cross-section of countries and found that the relation between income and deforestation resembled a U-shaped parabola in Asia and an inverted U-curve in Latin America and Africa. Using more recent information, Bawa and Dayanandan (1997) arrived at roughly the same conclusion.

Crop Yields

The evidence from household, village, and regional studies suggests that crop yields have mixed effects on forest clearance. Cleaver and Schreiber (1991) and Southgate (1991) found evidence in Africa, Brazil, and Ecuador that modern farm technologies reduced deforestation.

Using longitudinal information (1970-1982) from about 250 villages and 4,527 households in India, Foster, Rosenzweig, and Behrman (1998) discovered that doubling crop yields increased deforestation by about six percent. They also found that the presence of a factory in a village reduced the amount of deforested area by 19 percent. Because improvements in farm productivity encourage the growth of economic activities unrelated to farming (see chapter 2, "Comparing Approaches"), one could argue that technological progress in Indian agriculture enhanced conservation—if one takes into account both the direct and indirect effects of improved farm productivity

(Rosenzweig, Behrman, and Vashishtha 1995) and one does not limit oneself to estimating only the direct effect of crop yields on the area of forest cleared.

Private Time Preference

Few researchers have estimated the effects of private time preference on conservation (Cuesta, Carlson, and Lutz 1997). In the highlands of Ethiopia, researchers found that higher rates of private time preference induced farmers to lower investments in the control of soil erosion (Holden, Shiferaw, and Wilk 1998). Among the Tsimane´ Amerindians of the Maniqui River in the Bolivian rainforest, high rates of private time preference were associated with less clearance of forest (see chapter 12, "Time Preference, Markets, and the Evolution of Social Inequality"). Researchers carried out an experiment using real (rather than hypothetical) rewards of food to elicit private time preference. Researchers estimated the link between impatience and the area of old-growth forest cleared and found that patient people cleared more old-growth forest than impatient people.

Variables

The empirical evidence used to test the hypotheses came from household surveys done among the Tawahka, Mojeño, Yuracaré, Chiquitano, and Tsimane´. Table 5-1a and 5-1b contains definition and summary statistics of the variables used in the analysis. The measurement of the variables is discussed next.

Dependent Variable

The area of old-growth forest cleared the year before the interview was used as a dependent variable. Except for the Tawahka, where fields were measured, estimates of area cleared by other groups were obtained from interviews. Since the dependent variable was censored at zero (except among the Tsimane´ of the river Sécure) a tobit regression was used. Thirty-seven percent of the households interviewed did not cut old-growth rainforest. The share of households that did not cut old-growth forest ranged from

TABLE 5-1a *Definition and Summary Statistics of Variable for Tawahka*

Variable	Definition	Tawahka		
		Obs	Mean	Sd
OGF	Ha. of old-growth forest cut	98	.53	.77
Fallow	Ha. of fallow forest cut	98	.95	.95
Education	Maximum education of household head	101	2.49	2.44
HHsize	Total number of people in house	101	7.75	3.34
Resdur	Number of continuous years of residence in village	100	20.06	15.01
Wealth	Value of chickens, pigs, and cattle*	101	2.01	1.57
Farm income	Imputed farm income=quantity of principal annual crops (e.g., maize, rice, peanuts) multiplied by village price	101	6841	5889
Labor income	Wage income outside the village; for Tawahka=share of income from wage labor	90	.47	.44
Rice yield	Rice yields; *arrobas/tarea*; 4 *arrobas*=100 lbs; 10 *tareas*=1 hectare in Honduras	96	11.43	7.94
Maize yield	Same as rice yields (riceyld)	81	7.62	6.01
Illness	Number of days ill during bean season, 1994	97	13	25
Time pref	Private time preference in logarithms. For definition and computation, see Godoy and Kirby (2000).	n/a	n/a	n/a

NOTES   *Value is in lempiras (U.S.$1=9.40 La). "n./a" means not available.

a low of 25 percent among the Tsimane´ to a high of 50-56 percent among the Mojeño and the Yuracaré.

Explanatory Variables Besides Crop Yields and Private Time Preference

If one models forest clearance as consumption, then clearance should reflect demographic attributes of the household, life-cycle and human-capital attributes of the household heads, and the type and level of income of the household. Household size was included for demographic characteristics. Household size should bear a positive relation to forest clearance because it increases the demand for food and the amount of laborers available to clear

TABLE 5-1b Summary Statistics of Variables for Mojeño, Yuracaré, Tsimane´, and Chiquitano

Variable	Mojeño			Yuracaré			Tsimane´			Chiquitano		
	Obs	Mean	Sd	Obs	Mean	Sd	Obs	Mean	Sd	Obs	Mean	Sd
OGF	131	.41	.49	62	.40	.58	237	.63	.74	240	.58	.54
Fallow	132	.58	.58	62	.40	.40	237	.58	.79	240	.69	.47
Education	132	2.35	1.70	62	1.72	1.65	237	1.21	1.76	240	3.64	2.56
HHsize	132	6.43	2.27	62	6.35	2.47	237	5.01	2.99	240	6.16	2.59
Resdur	132	13.3	12.8	62	16.4	12.7	234	17.5	15.9	240	32	14.8
Wealth*	132	3968	3918	62	3760	2516	236	1889	1792	239	4553	4781
Farm income	132	1072	879	62	1353	1387	234	1019	1106	240	1394	1219
Labor income	132	687	1469	62	498	1086	237	1076	2352	240	1895	3388
Rice yield	119	8.13	5.22	56	9.23	5.55	208	12.5	14.7	186	6.59	6.05
Maize yield	106	5.73	3.47	52	5.23	2.89	195	9.17	13.4	222	8.09	6.41
Illness**	132	1.31	3.5	62	.48	1.45	236	6.16	10.7	237	2.13	4.73
Time pref	132	.06	.09	62	.05	.09	68	.04	.08	242	.1	.1

* value of 13 physical assets for Mojeño, Yuracaré, and Chiquitano in Bolivianos (U.S.$1=5.23 Bol, 97'); for Tsimane´ in Bolivianos (U.S.$1=5.05 Bol, 96').

** number of days ill during forest-cutting season in Bolivia

NOTES   Tsimane´ includes the population in Territorio Uno surveyed in 1996 (n=209) and the Tsimane´ of the river Sécure (n=29) surveyed in 1998. For the Chiquitano and for the groups in the river Sécure, only one estimate of time preference from one of the two randomly-chosen heads of households was used. Among Tsimane´, estimates of time preference for both household heads were used.

the forest, although the effect of demographic variables may wane as villagers tighten their links to the market (see chapter 7, "Chayanov and Netting: When Does Demography Matter?"). The characteristics of the household head included education and duration of residence in the village. Education has been shown to deter forest clearance in several household studies and international comparisons (Godoy and Contreras 2001). Residence duration in a village proxies for security of tenure (Godoy, Kirby, and Wilkie 2001).

This study controls for wealth—measured by the value of selected physical assets. Wealth should make it easier for households to overcome borrowing constraints and acquire resources to clear more forest and buy improved farm technologies. Dummy variables for ethnic groups were included in the regression with the pooled sample.

Crop Yields and Private Time Preference

The yields of only two crops were included—rice (a traded crop) and maize (a less traded staple)—for two reasons. First, the two crops were found in all the cultures and allow one to compare results and draw generalizations from the comparison. Some crops are important, but only to some cultures. Peanuts, for instance, are prominent among the Chiquitano, but beans and cacao are prominent among the Tawahka. Focusing on peanuts, beans, and cacao would have allowed us to enrich the story of the Tawahka and Chiquitano, but at the expense of being able to produce generalizations.

Second, the two crops represent a tradable cash crop (rice) and a non-tradable subsistence crop (maize)—allowing one to test the fourth hypothesis. Information on area planted and crop production was obtained from interviews rather than from direct measurements.

To elicit private time preference, subjects were given nine, non-trivial choices between having money now or having more money in the future (Godoy, Kirby, and Wilkie 2001). Based on their responses, a discount rate consistent with their choices was estimated. To make the experiment real rather than hypothetical, one of the nine answers was selected at random and subjects were rewarded with the amount specified in that answer.

Logarithms of income, wealth, distance from village to the nearest market town, private time preference, yields, and area of forest cleared were taken to make the interpretation of results easier. When dependent and explanatory variables are expressed in logarithms, the estimated coefficients can be read as *elasticities*—or as the percentage of change in the dependent variable (in this case area of forest cleared) from a percentage of change in an explanatory variable.

Biases

Two potential biases deserve attention before turning to a discussion of the results—bias from omitted variables and bias from endogeneity. We did not collect information on some unseen attributes of localities, such as soil quality. If soil quality is positively associated with yields and income and the presence of old-growth forest, then failure to control for this unseen ecological attribute will bias the estimated coefficients of yields and income upwards. Village dummies were not used to correct for the bias because a village dummy would not have controlled for variation in soil qualities within the village.

Biases could also arise from endogeneity. Yields may affect forest clearance, but forest clearance may allow households to increase their income and improve their crop yields. To correct for the potential bias, a sensitivity analysis was done using illness as an instrumental variable for income and two-stage least squares.

Results

The regression results are presented in four tables. Table 5-2 contains regressions with farm and labor income measured separately to test whether the sign of the coefficient of farm income is positive and whether the sign of the coefficient of wage income is negative—as stated in Hypotheses 1-2. Table 5-3 uses the regression of table 5-2 but replaces the variables for imputed farm income and wage income with their sum (or total income) and includes a square term for income to test for non-linearities (Hypothesis 3). Table 5-4 replaces the variable income with two variables: rice yields and maize yields (Hypothesis 4). Last, the regression of table 5-5 uses the model of table 5-3 but adds the variable for private time preference. Table 5-5 only includes the results for the groups in Bolivia, because information on private time preference among the Tawahka was not collected.

Hypotheses 1-3: Forest Clearance and Income

The information in table 5-2 supports the hypothesis that income from annual crops bears a positive relation to the cleared area of old-growth forest. Except for the Tawahka, all results were statistically significant at the 95 percent confidence level or above. In the pooled sample, doubling income from annual crops increased the area cleared of old-growth forest by about 44 percent (t=6.51; p>|t|=0.01%).

TABLE 5-2 *The Role of Wage Income and Farm Income on Deforestation*

Variable	Pooled Coef	Se	Tawahka Coef	Se	Tsimane′ Coef	Se	Mojeño Coef	Se	Yuracaré Coef	Se	Chiquitano Coef	Se
Education	$-.52^3$	.08	$-.69^2$	.31	$-.91^3$	.15	$-.73^2$	.34	-.65	.58	$-.20^1$	.12
HHsize	.01	.06	.002	.25	.07	.07	.19	.25	.56	.40	-.03	.11
Resdur	$-.04^3$	.01	$-.1^1$	.05	-.02	.01	-.07	.04	.02	.07	$-.03^1$	.02
Wealth	$-.38^3$	.11	-.27	.22	$-.24^1$	.15	.33	.60	-1.52	1.02	-.02	.28
Farm income	$.44^3$	.06	.39	.25	$.44^3$	.08	$.46^2$	.23	3.09^3	1.18	$.45^3$	.11
Labor income	.02	.03	-.42	.34	-.01	.04	$.17^1$	.1	.14	.17	.08	.06
Distance	-.21	.22	-.49	.63	.05	.22	-1.57	.99	-3.41	2.22	-.45	.56

Observations

Left censored	282	44	57	66	35	80
Right censored	469	44	174	65	27	159
Total	751	80	231	131	62	239

NOTES   Dependent variable is logarithm of hectares of old-growth forest cut. Regressions are tobits through origin; regression for pooled sample includes dummies for ethnic groups. Area of forest cut, income, wealth, and distance expressed in logarithms. 1, 2, and 3 are significant at ≤10%, ≤5%, and ≤1%.

The information in table 5-2 did not support the hypothesis that wage labor reduces the clearance of old-growth forest. For the pooled sample, and for the Yuracaré, Mojeño, and Chiquitano, an increase in income from wage labor was associated with an increase in the area of old-growth forest cut. Only among the Mojeño was the relation statistically significant. Among the Mojeño, a doubling of income from jobs outside the farm was associated with an increase in the area of old-growth forest cut by about 17 percent (p>|t|=10.1%). Only with the Tawahka and the Tsimane′ did one find the hypothesized negative link between wage income and forest clearance, but in neither group was the relation statistically significant at the 90 percent confidence level or above.

Support for the third hypothesis—non-linear links between total income and forest clearance—was present but weak, as shown in table 5-3. First, one

TABLE 5-3 The Role of Total Income on Deforestation and Test of Non-linearities

	Pooled		Tawahka		Tsimane′		Mojeño		Yuracaré		Chiquitano	
Variable	Coef	Se	Coef	Se	Coef	Se	Coef	Se	Coef	Se	Coef	Se
Education	$-.56^3$	.09	$-.69^2$	.30	-1.10^3	.16	$-.87^3$	.34	$-.93^1$	.56	-.13	.13
HHsize	.01	.06	-.05	.26	.02	.07	.26	.26	.60	.43	-.001	.12
Resdur	$-.04^3$	.01	$-.13^2$	.05	-.01	.01	-.06	.04	.04	.07	-.03	.02
Wealth	$-.24^2$	.11	$-.36^1$	.22	-.01	.16	.33	.58	-1.05	.99	.22	.28
Income	$.23^1$	.12	$.80^2$	.36	.34	.25	1.55	1.10	5.68^3	1.90	.37	.26
Income2	-.008	.006	-.001	.001	-.002	.002	-.001	.001	$-.006^3$	.002	-.002	.001
Distance	.17	.21	-.07	.58	.27	.24	-1.4	.99	-1.18	1.58	-.46	.58
Observations												
Left censored	282		44		57		66		35		80	
Right censored	469		44		174		65		27		159	
Total	751		80		231		131		62		239	
Test (Prob>F)												
%	13.54		1.98		39.54		3.16		1.47		34.24	

NOTES   Same notes as in table 5.2, except income and income squared expressed in levels ('000s). F test is for joint statistical significance of income and income squared. 1, 2, and 3 are significant at $\leq$10%, $\leq$5%, and $\leq$1%.

might have expected to find different forms of non-linearities. Only the inverted, U-shaped parabola emerged from the analysis. In the five cultures, clearance of old-growth forest increased at low levels of income, but then decreased once households passed an income threshold. Second, the relation between deforestation and total income was statistically significant at a 90 percent or higher confidence level only among the Tawahka (prob>F=1.98%), Mojeño (prob>F=3.16%), and Yuracaré (prob>F=1.47%), but not among the Tsimane′ or Chiquitano.

Hypothesis 4: Forest Clearance and Crop Yields

Contrary to expectations, the yields of maize and rice did not seem to affect the amount of old-growth forest cut in the way predicted. The result presented in table 5-4 suggests that improvements in the yield of either crop were associated with an increase (rather than a decrease) in the area of old-growth forest cleared in all groups except among the Chiquitano.

In no group was the result statistically significant at the 90 percent confidence level or above. The results also suggest that yields had only a small physical impact on forest clearance. The results for the pooled sample, for example, suggest that doubling yields of rice and maize were associated with an increase in deforestation of only about ten percent. Although ten percent

TABLE 5-4 *The Role of Crop Yields in Deforestation*

	Pooled		Tawahka		Tsimane´		Mojeño		Yuracaré		Chiquitano	
Variable	Coef	Se	Coef	Se	Coef	Se	Coef	Se	Coef	Se	Coef	Se
Education	-.28[3]	.08	-.35	.31	-.63[3]	.15	-.58[1]	.35	-.26	.57	-.002	.13
HHsize	.03	.06	.006	.27	.06	.06	.02	.27	.45	.44	.05	.12
Resdur	-.03[3]	.01	-.06	.05	-.02[1]	.01	-.06	.05	.01	.07	-.02	.02
Wealth	-.09	.11	-.06	.26	.06	.22	-.07	.65	-1.37	1.07	.01	.29
Rice yield	.05	.11	.19	.71	.19	.19	-.13	.58	2.72	1.70	-.02	.13
Maize yield	.05	.13	.17	.64	.13	.15	.19	.58	-.06	.70	-.02	.21
Distance	-.13	.22	-.36	.72	-.21	.28	-.04	1.08	-.18	1.81	.10	.61

Observations

	Pooled	Tawahka	Tsimane´	Mojeño	Yuracaré	Chiquitano
Left censored	183	38	29	45	24	47
Right censored	403	40	152	54	24	133
Total	586	78	181	99	48	180

Test (Prob>F)

	Pooled	Tawahka	Tsimane´	Mojeño	Yuracaré	Chiquitano
%	55.77	86.58	29.25	94.67	28.41	96.46

NOTES Same notes as in table 5.2, except rice and maize yields (in logarithms) replace farm and wage income. F test is for joint statistical significance of maize and rice yields.

is more than the six percent rate estimated by Foster and his colleagues in India (Foster, Rosenzweig, and Behrman 1998), the level is low by absolute standards. Among the Chiquitano, a doubling of yields lowered the amount of old-growth forest cleared by only about four percent—the result was statistically insignificant.

Hypothesis 5: Forest Clearance and Private Time Preference

The results presented in table 5-5 suggest that private time preference has an ambiguous and statistically insignificant impact on the area of cleared forest.

TABLE 5-5 *The Role of Time Preference in Deforestation*

Variable	Pooled		Mojeño		Yuracaré		Tsimane′		Chiquitano	
	Coef	Se	Coef	Se	Coef	Se	Coef	Se	Coef	Se
Education	$-.18^2$	.09	$-.66^3$	.25	-.86	.84	-.30	.44	-.02	.11
HHsize	.06	.09	.17	.25	.54	.55	.04	.07	-.002	.13
Resdur	-.02	.01	.0008	.04	.06	.10	-.008	.01	-.03	.02
Wealth	-.34	.23	-1.36^3	.57	-1.28	1.27	.10	.13	-.30	.31
Income	$.55^2$	.29	2.44^1	1.45	3.75^1	2.19	.36	.84	.18	.22
Income2	-.03	.02	-.18	.22	-.42	.27	.13	.33	-.01	.01
Time pref	-.06	.08	.10	.18	-.21	.44	-.00008	.05	-.13	.12
Distance	2.50	1.40	1.03	.83	-1.13	2.09	-.08	.19	.44	.63
Observations										
Left censored	132		45		25				61	
Right censored	249		56		16				124	
Total	381		101		41		54		185	

NOTES   Same notes and regression as in table 5.3, except measure of private time preference added if consistency in time preference is above 0.67. For definition of consistency, see Godoy, Kirby, and Wilkie (2001). Regression for Tsimane′ is ordinary least square because dependent variable is uncensored.

Among all groups except the Mojeño, higher rates of private time preference were associated with smaller areas of cleared forest. In the pooled sample, a doubling in the rate of private time preference reduced the cleared area of old-growth forest by about 6 percent (p>|t|=78.11%). In none of the four groups were results statistically significant. Among the Tsimane´ the effect of private time preference on the area cleared was physically trivial.

Sensitivity Analysis and Controlling for Reverse Causality

To test whether the results just discussed also apply to the cutting of secondary-growth forest, the regressions of tables 5-2 through 5-5 were re-run with the area of secondary-growth forest cleared as a dependent variable. The results (not shown here) were essentially the same. Income from annual crops increased the area of cleared secondary-growth forest, but income from wage labor still had statistically weak and ambiguous effects on forest clearance. The relation between total income and the cleared area of fallow forest resembled an inverted, U-shaped curve in all cultures except the Chiquitano. Only among the Mojeño and the Yuracaré was the relation statistically significant at the 95 percent confidence level or above. Yields also had positive and statistically insignificant effects on the clearance of secondary-growth forest, except among the Tsimane´. There, a doubling of rice yields was associated with 37 percent greater clearance of secondary-growth forest (p>|t|=0.1%). Last, private time preference bore a positive—but statistically insignificant—relation to the cleared area of fallow forest in all cultures. In summary, crop yields, private time preference, and farm, labor, and total income seemed to affect the clearance of old and secondary-growth forest in similar ways.

Since income (particularly farm income) is endogenous, the regressions of table 5-3 were re-run for the pooled sample, with illness used as an instrument for income. The coefficient of income switched signs and became statistically insignificant when using two stage least squares, suggesting that some of the results discussed in this chapter must be read with care, since they may be sensitive to endogeneity.

Conclusion

Despite problems of endogeneity, the study contains several tentative lessons for anthropologists, policy-makers, and conservationists.

First, the results suggest that markets worsen conservation before improving it. Greater integration into the market through the production of annual crops is associated with greater deforestation. After households pass a threshold of income they cut less forest. The reasons for the decline in deforestation beyond a certain threshold of income are unclear. The evidence presented suggests that forest clearance does not decline because indigenous people increase their work outside of the farm or because they have higher crop yields. Forest clearance also does not decline because economic development, by raising wealth and income, lowers people's private time preference. Private time preference has no discernable effect on how much forest people cleared. Deforestation also does not decrease because villages have no more forest left. The regressions of table 5-3 were re-run for the groups in Bolivia with area of available forest (proxied by the time it took to reach the closest old-growth forest from the center of the village) as an additional explanatory variable. There was still statistically strong evidence for an inverted, U-shaped parabola.

Second, material determinants (embodied in such things as imputed farm income) seem to overshadow cognitive variables (embodied in private time preference) in explaining forest clearance. This lends support to the materialist interpretation of culture—at least on this topic.

Last, some variables amenable to manipulation by policy-makers seem to have more potential for curbing deforestation than others. The results presented do not support the idea that greater off-farm employment or improvements in crop yields reduce deforestation—at least not in the short run or for the populations under study.

The results presented, however, do suggest that improvements in education might be associated with lower rates of deforestation. The information in tables 5-2 through 5-5 suggests that schooling reduces the amount of old-growth forest cleared by all groups, regardless of the econometric model used (Godoy 1994, 1999). In the pooled sample and each culture, formal schooling was associated with a smaller area of cleared forest. Those results were statistically significant at the 90 percent confidence level or above. Elsewhere, the positive environmental externality produced by one more year of schooling (Godoy and Contreras 2001) was estimated. This might justify a schooling subsidy as a progressive and fiscally responsible way of simultaneously improving the human capital of lowland Amerindians and enhancing conservation.

Game Consumption, Income, and Prices: Empirical Estimates and Implications for Conservation

For many years, anthropologists and ecologists have been discussing the link between economic development and changes in the way lowland Amerindians hunt. So far, however, they have provided few direct, quantitative estimates of how income and prices affect the consumption of game (Jorgenson 1997; Stearman 1990; Vickers 1988, 1994). Such estimates are useful in identifying the wildlife most likely to face hunting pressure and extinction as rural economies modernize. Such estimates are also useful in identifying broader policies that are likely to improve conservation. This chapter examines how changes in the income of indigenous people and the price of wildlife (and its substitutes) affect the consumption of game, and how economic development might affect the availability of wildlife.

Economic development could affect wildlife through at least two channels:

1. changes in consumption, and

2. changes in the costs of hunting.

Changes in consumption could arise from changes in income or from changes in the price of substitutes for game meat. Assuming game meat is like most other foods, higher income will probably increase consumption of game meat. Economic development could also deflect consumption away from game meat by lowering the price of substitutes—such as the price of meat from domesticated animals. Commercial production of poultry, beef, fish, and pork would increase the supply of animal proteins and deflect consumption away from bush meat by lowering the price of meat from domesticated animals.

Changes in consumption do not only arise from changes in demand. They can also arise from changes in the costs of extraction. The introduction of new hunting technologies or the opening of new employment opportunities in the countryside could affect foraging costs. Firearms produce more kilograms of bush meat for each hour of effort than traditional hunting technologies. By lowering the costs of foraging and game meat, new hunting technologies increase game consumption. Changes in the costs of extraction could also come from more jobs outside of the farm. Increased employment opportunities raises the value of time of foragers, making them less likely to hunt (or to hunt more selectively) for primarily valuable or larger game. As in the case of demand, so too in the case of supply—economic development could produce opposite effects on consumption by simultaneously lowering and raising the costs of foraging through the use of more efficient foraging technologies (which would lower foraging costs) and through the creation of more and higher-paying jobs (which would raise foraging costs).

This chapter focuses on the demand side of how economic development affects the consumption of edible wildlife. How changes in the costs of hunting affect consumption of wildlife are not analyzed. This chapter instead presents one of the first empirical estimates of how changes in income, the price of game meat, and the price of substitutes for game meat affect the consumption of wildlife among indigenous people. It then explores how economic development may affect the availability of game in the forest.

To estimate the effect of economic development on game consumption the study draws on 1998 survey information collected from the Yuracaré, Chiquitano, Mojeño, and Tsimane' of Bolivia. The estimates from the survey are then compared with estimates from the panel information of the Tawahka. Direct observations from repeated census of animals in the Tawahka villages of Krausirpe and Yapuwás were used to estimate the effect of economic development on the availability of game.

The Role of Income and Prices in Game Consumption: Implications for Conservation

This section draws on price theory to explain how changes in income, the price of bush meat, and the price of substitutes for bush meat might affect the consumption of wildlife. The next section provides empirical estimates for some of the relationships.

Income

A description of the relationship between income and game consumption draws on the concept of an income elasticity of consumption—defined as the percent change in consumption from a percent change in income. All else held constant, an increase in income could produce three changes in the consumption of wildlife, depending on whether wildlife is an inferior good, a superior good, or a necessity. Superior animals are animals whose consumption increases by more than one percent for every percent increase in income. Necessities are animals whose consumption increases by less than one percent for every percent increase in income. Inferior animals are those whose consumption falls when incomes rise. Normal goods are goods with a positive income elasticity of consumption, and include necessities and superior goods. An animal may fall under more than one category, depending on the level of income of the consumer. Among poor people, for example, an increase in income may at first induce an increase in game consumption. Beyond a threshold of income, game consumption may grow more slowly or perhaps even fall. The words superior, normal, necessities, and inferior summarize an empirical relationship between the consumption of an animal or a group of animals and income—the words do not imply that animals are better or worse than each other.

Prices

On the demand side, two prices will drive the consumption of game—the price or value of game and the price or value of close substitutes. All else held constant, an increase in the price of game will decrease the consumption of game. We refer to this relationship as the own-price elasticity of consumption—defined as the percent change in the consumption of an animal brought about by a corresponding change in the price of that animal. The higher the own-price elasticity of consumption, the greater the number of substitutes available to consumers, since a small change in the price of the meat from that animal will produce a large change in the quantity consumed.

An increase in the price of another source of animal protein besides game meat—such as poultry, beef, or pork—ought to increase game consumption if game meat and domesticated meat are substitutes for each other. An increase in the price of domesticated meat ought to decrease consumption of game if the two goods complement each other. The relationship between a good and its substitutes or complements is referred to as the cross-price elasticity of consumption—defined, in this case, as a change in consumption of game meat from an animal arising from a change in the price of another source of animal protein (both in percentages). A negative cross-price elasticity of consumption (between meat from game and meat from domesticated animals) indicates that the two goods are complements. A positive cross-price elasticity of consumption indicates that the two goods are substitutes. A high cross-price elasticity of consumption between game meat and meat from domesticated animals indicates that there is potential to reduce the pressure on wildlife through the development of cheaper sources of animal protein.

Ecological Implications

The effects of economic development on the availability of wildlife will be ambiguous, even if one limits the story to the demand side and even if one assumes that all game are normal goods. By increasing income, economic development will increase the consumption of most types of game and, in so doing, will reduce the availability of wildlife. Economic development will also increase the size of the human population, at least in the short run, and increase the demand for game meat and the corresponding pressure on wildlife. An increase in income and population will work together to undermine the conservation of wildlife.

Economic development, however, will also lower the price of game meat substitutes through commercial production of poultry, beef, fish, and pork—reducing demand for game and hunting pressure. The net effect of economic development on game consumption and availability is ambiguous, since economic development will cause demand to shift in opposite directions.

Results become even more ambiguous if one introduces changes on the supply side that often accompany economic development. Such changes include the introduction of more efficient hunting technologies or the creation of non-agricultural jobs. The changes pull consumption of wildlife in opposite directions from the supply side. Increased hunting pressure from changes in technology or changes in demand could also cause some r-selected or fast-reproducing animals to increase their rate of reproduction as they try to adapt to more predation (Wilson and MacArthur 1967). Because the effects of economic development on game consumption and the availability of wildlife are theoretically unclear, they must be estimated through empirical analysis.

Goals, Variables, and Econometric Models

The empirical analysis below presents estimates of: 1) the own-price, cross-price, and income elasticity of consumption for fish and all other game among the four groups in Bolivia, and 2) the effect of economic development on the availability of game in two Tawahka villages—representing different points in the autarky-to-market continuum.

Variables

The dependent variables are the reported kilograms of fish and game meat brought to the household during the week before the interview. Seventy-three percent of the interviews were done during the rainy season (between February and April, 1998). Weekly consumption per person of game and fish from the sample of 483 households averaged 1.08 kilograms of game meat (standard deviation=3.29) and 2.43 kilograms of fish (standard deviation=5.05). A third variable was created by adding the quantities of fish and game meat to capture the total weight of wildlife consumed each week by each person. Explanatory variables included total income per person, wealth per person, total household size, education of male household head, village prices for fish and domesticated animals (chickens, ducks, pigs, and cattle), and dummy variables for villages and ethnic groups. Income included the following:

— imputed farm income from the harvest of maize, rice, and peanuts

— cash income from the sale of farm products and forest goods (excluding game)

— cash income from wage labor

Consumption, income, wealth, household size, education, and prices were transformed into logarithms. Ordinary least squares with robust standard errors were used to estimate elasticities.

Results

Results are presented in four sections. The first three sections present the estimates of the income, own-price, and cross-price elasticities of consumption (summarized in table 6.1) for all animals, game, and fish and for the top and bottom half of the income distribution. There were not enough observations to do the analyses by income quartiles. The estimates allow one to decide if wildlife are normal or inferior goods, and explore how elasticities of consumption for different types of animals change with income. The fourth section explores whether the forest of the much larger and richer village of Krausirpe has less game than the poorer and much smaller village of Yapuwás. The estimate allows one to test whether economic development reduces the availability of edible game in the forest.

Income Elasticities

The results shown in table 6-1 suggest that fish and other game, taken together, seem to be an inferior good. The income elasticity of consumption for all game in the pooled sample was -0.20 (p>|t|=3.7%), suggesting that a doubling of income would reduce consumption of wildlife by about 20 percent. An increase in income would seem to have a greater effect in curbing consumption of wildlife in the bottom half of the income distribution than it would in the top half. The income elasticities of consumption in the bottom and top halves were -0.27 (p>|t|=2.5%) and -0.19 (p>|t|=55.5%).

These results gloss over differences between different types of animals. When the analysis is done separately for fish and game meat, a different story emerges. Fish appear to be an inferior good, with an income elasticity of consumption for the pooled sample of -0.16 (p>|t|=4.6%). An increase in

TABLE 6-1 *Income and Own and Cross-Price Elasticities of Game and Fish Consumption: Bolivian Lowlands*

	Pooled	Bottom	Top
All (game and fish)			
Income	$-.20^2$	$-.27^2$	-.19
Own	-4.11^3	-4.54^3	-2.10
Cross	-1.54	5.70^3	Dropped
R2	.56	.63	.53
Obs	461	230	231
Fish			
Income	$-.16^2$	$-.28^3$	-.01
Own	-4.59^3	-4.02^3	-1.98
Cross	1.70^3	-2.55^2	Dropped
R2	.73	.74	.76
Obs	461	230	231
Game			
Income	.12	.20	-.11
Own	-2.88^3	-3.65^3	-1.82
Cross	1.22	8.18^3	Dropped
R2	.30	.46	.26
Obs	461	230	231

NOTES Numbers reported are the income, own-price, and cross-price elasticities of consumption. Elasticities come from ordinary least squares regressions with Huber White robust standard errors and constant. Besides logarithms of income, price of fish, price of domesticated animals, and of wealth, regressions also include logarithms of education of male head and household size, and dummies for ethnic groups and for villages. Column called "Bottom" refers to households with less than the median income ($\leq$1877 *Bolivianos*/household); "Top" refers to households with more than the median income. 1, 2, and 3 significant at $\leq$10%, $\leq$5%, and $\leq$1%.

income seems to have a much stronger effect in curbing fish consumption in the bottom half of the income distribution (elasticity -0.28; p>ltl=0.5%) than in the top half (elasticity -0.01; p>ltl=92.8%). On the other hand, game meat appears to be a necessity in the pooled sample (elasticity 0.12; p>ltl=19.1%) and in the bottom half of the income distribution (elasticity 0.20; p>ltl14.6%). It seems to become an inferior good, however, in the top half (elasticity -0.11; p>ltl=75.7%). Since the income elasticities of consumption for game meat hover around zero and are statistically insignificant at the 90 percent confidence level or above, one might tentatively conclude that game meat is a necessity bordering on being an inferior good.

The evidence suggests that growth in income probably discourages the consumption of fish and increases the consumption of game meat, but at low rates. To further probe this point, the income elasticities of consumption from Bolivia were compared with the income elasticities of consumption from the Tawahka. The results of random-effect estimations for the pooled sample of Tawahka, adjusted by adult-equivalents living in the household, suggests that the income elasticity of consumption for fish was close to zero and statistically insignificant (-0.01, p>lzl=84.9%). As in Bolivia, game meat was a normal good in the pooled sample (elasticity of 0.19, p>lzl=8.9%) and in the bottom half of the income distribution (elasticity of 0.50, p>lzl=0.3%), but appears to become an inferior good in the top half (elasticity of -0.6, p>lzl=74.1%).

Own-Price Elasticities

Although information on the village price of game meat was not collected, information on the village price of fish was. The only own-price elasticity that could be estimated with accuracy was the own-price elasticity for the consumption of fish. If one assumes that the price of fish and the price of game meat move in unison, then the price of fish could be used as a proxy for the price of game meat—which is the assumption made in estimating the own-price elasticity of consumption for game meat. Care should be taken in reading the row called *own* for game meat in table 6-1 because those estimates refer to changes in the consumption of game meat produced by a change in the price of fish, not game.

Bearing the caveat in mind, one can infer from table 6-1 that wildlife has an elastic demand. Consumption appears to be almost twice as elastic in the bottom half of the income distribution than in the top half for all animals. The own-price elasticity of consumption for fish in the bottom half of the income distribution, for example, was -4.02 (p>ltl=0.1%), but it was only -1.98 (p>ltl=16.4%) in the top half. The high own-price elasticity of consumption suggests that indigenous people may have many sources of animal protein available to them—a finding with positive and with negative implications for conservation, as discussed in the conclusion of this chapter.

Cross-Price Elasticities

Because of multi-collinearity, the cross-price elasticities of consumption of game for the top half of the income distribution were not estimated. The results for the bottom half, however, suggest that fish is a complement and game meat is a substitute for domesticated animal meat. An increase in the price of domesticated animals reduces consumption of fish, but increases consumption of game meat in the bottom half of the income distribution. A one percent increase in the price of domesticated animals results in a 2.55 percent (p>|t|=5.3%) decrease in fish consumption, but a large increase in game consumption (elasticity 8.18; p>|t|=0.1%).

Comparing Availability of Game in Rich and Poor Communities

In the poor and rich villages of Yapuwás and Krausirpe, researchers encountered a total of 28 species, principally birds (n=13) and mammals (n-9), followed by monkeys (n=3), carnivores (n=2), and reptiles (n=1). Of the 28 species, 16 species were encountered with more frequency in Yapuwás and 10 species were encountered with more frequency in Krausirpe. Two species showed the same frequency of encounters in the two sites (Demmer et al. 2001).

A bivariate test comparing the availability of animals between the two sites suggests that the frequency of encounters for peccaries, howler monkeys, and six types of birds was higher (and statistically significant) in the poorer village of Yapuwás. Krausirpe had a higher (and statistically significant) rate of encounters for agoutis. Very few of the animal species showed statistically significant differences in encounter rates between the sites after controlling for the role of third variables.

A random-effect probit model was used to predict the likelihood of sighting an r-selected species while controlling for the role of precipitation, vegetation type in the transect at the place of sighting, season, and using a dummy variable for the village. The village dummy picks up the effect of economic develoment and all the unseen, fixed, ecological, and climatological variables of the communities that influence the abundance of game but are not captured by the ecological and climatological variables measured (e.g., vegetation type in transect, precipitation). The results of the analysis (not shown) showed that being in the poorer village of Yapuwás had a negative and statistically insignificant effect on the likelihood of sighting an animal.

One can speculate about why the availability of game might not differ in statistically significant ways between two communities that have such large differences in levels of economic development. First, perhaps the wildlife in the area is sensitive to anthropogenic pressure, so even modest levels of economic development might bring the animal availability of the two sites down to a lower common denominator. Second, perhaps economic development affects wildlife—but only after economies pass a threshold, which the Tawahka have yet to reach. One might be able to answer these two points with information from non-hunted areas, which was not collected. Last, perhaps economic development undermines the availability of game. A small sample size of encounters collected over only 2.5 years of research might not provide enough observations to answer the query in a convincing way.

Conclusion

Several tentative lessons for conservation emerge from the information and analysis presented. Some of the lessons bode well for conservation, but others do not.

On the positive side, economic development will probably enhance conservation by improving income and lowering the price of domesticated animals. The results of the analysis suggest that growth in income will reduce consumption of fish for the poor and rich alike, and will likely reduce consumption of game meat once people pass a threshold of income. Lowering the price of domesticated animals would seem to have a large effect in reducing the consumption of game meat among the bottom half of the income distribution.

The results of the analysis, however, also show how economic development could worsen the conservation of wildlife. If the own-price elasticities of consumption are as high as these estimates suggest, changes in the marginal costs of hunting might have large effects on the amount of wildlife extracted from the forest. Higher wages and more jobs will cause people to consume much less wildlife, but the introduction of more efficient hunting technologies could have the opposite effect.

As discussed at the outset of the chapter, economic development produces so many opposing changes on the supply and demand sides of wildlife that it becomes difficult to predict the net effect of economic development on the availability of wildlife. One way to assess the effect is to compare the availability of game in villages that resemble each other in ecology and a broad range of cultural and socioeconomic attributes other than income. Such a controlled comparison is possible, but would require that a large number of villages be compared well and yield accurate estimates. The preliminary results of the more modest comparative study in only two Tawahka villages suggests that large differences in human population size and income seem to have no discernable effect in the abundance of game in the forest, after controlling for third variables.

The conclusion that economic development may have negligible effects on wildlife conservation echoes the finding of anthropologist William Vickers, who has done longitudinal research among the Siona-Secoya Indians in the Ecuadorian Amazon. In a study spanning more than two decades, Vickers found that:

> Despite 20 years of hunting pressure...and 12 years of increasing colonization, hunters were still finding tapirs, white-lipped and collared peccaries, woolly and howler monkeys, capybaras, agoutis, guans, and deer. The one animal that was reported to be absent was the curassow...most men are hunting less because of their greater involvement in cash-earning activities rather than because of game depletion per se.
>
> (Vickers 1994:330)

The comparative study of the determinants of wildlife consumption presented in this chapter, and the comparative evidence of the clearance of old-growth forest presented in chapter 5, "Forest Clearance: Income, Technology, and Private Time Preference," both suggest that economic development may not undermine the resource base of indigenous people—at least, it may not do so in a linear way, or as much as one might think.

Chapter 7

Chayanov and Netting:
When Does Demography Matter?

The last two chapters have tried to show that markets and economic development may not reduce the availability of wildlife and may even encourage the consumption of substitutes, such as meat from domesticated animals. They also suggested that markets might improve and worsen deforestation—depending on the household's type and level of income. This chapter moves away from exploring the effect of markets on indigenous people's use of natural resources and begins to explore how markets affect the role of demography in production.

The starting point is Alexander V. Chayanov's classic treatise, *The Theory of Peasant Economy,* first published in Russia in 1925 and translated into English in 1966. In that book, Chayanov said that for isolated Russian peasant households with access to ample land, the number of consumers relative to the number of producers dictated how hard people had to work. In relatively autarkic households with access to ample land,

> ...the force of the influence of consumer demands ...is so great that...the worker, under pressure from a growing consumer demand, develops his output in strict accordance with the growing number of consumers. The volume of the family's activity depends entirely on the number of consumers and not at all on the number of workers.

(Chayanov 1986:78, quoted in Netting 1993:305)

Chayanov's book lay dormant for five years after its translation into English, until Marshall Sahlins (1971, 1972) offered his own interpretation of Chayanov's book. Like other anthropologists since the appearance of the translation, Sahlins found much useful information in Chayanov's work. Chayanov's theory was meant to apply to non-industrial, non-market rural societies without land constraints and without a well-developed market for labor. These were the societies that anthropologists typically studied (Chibnik 1987).

Sahlins accepted Chayanov's idea that demography determined production in non-industrial rural economies, but he extended this line of reasoning to argue that the intensity of production in those societies also reflected kinship and political organization. Sahlins pointed out that in non-industrial rural economies, kinship, ideology, reciprocity, and political organization, or (more broadly) aspects of superstructure and structure, eclipsed economics and ecology in shaping production.

The idea that the composition of the household or the household developmental cycle—proxied by variables such as the ratio of consumers to adult producers—affects how hard rural people have to work reached iconic eminence after Sahlins' interpretation of Chayanov's work (Netting 1993:293). Inspired by the writings of Chayanov and Sahlins, many anthropologists have begun searching inside household demographics for the key to farm output.

But perhaps the key to output does not lie inside the household. Perhaps one is likely to find the key to output in household demographics only in some cases. In the concluding chapter of *Smallholders, Householders, Farm Families and the Ecology of Intensive, Sustainable Agriculture* (the last book he wrote before his untimely death), the great human ecologist Robert Netting offered a thoughtful meditation on Chayanov's work. With his customary thoroughness, keen sense of problem, and deep respect for facts, Netting reviewed the writings of many scholars who had analyzed how the household development cycle affected production in non-industrial economies. He found that Chayanov's theory that the ratio of consumers to adult workers drove production did not stand up well to the ethnographic facts. He found ethnographic examples in which the link between the intensity of production and household demographics was weak or ran counter to Chayanov's expectation. Netting realized that some researchers (e.g., Chibnik 1987; Minge-Kalman 1977; Tannenbaum 1984) had tested Chayanov's ideas using bivariate analysis. It was legitimate for him to wonder whether the results would also hold after controlling for the role of omitted variables such as population density, or whether they would hold in more complex rural economies than those studied by Chayanov.

Puzzled by the lack of a better fit between a simple idea and a messier ethnographic reality, Netting asked why Chayanov's model did so poorly. Drawing on the works of Ellis (1988) and Shanin (1986) before him, and some of the empirical work he had reviewed, Netting hypothesized that even in non-industrial rural economies farm output probably reflected more than household demographics—though he seemed unwilling to go as far as Sahlins in putting kinship, political organization, or ideology in the center stage. Chayanov had discussed the role of other variables besides demography, but he had relegated those variables to the background (Chibnik 1987:78). Netting listed some of the non-demographic variables that probably affected farm output—such as illness, crop failure, access to markets, and modern farm technologies—but he seemed more comfortable opening the debate than closing it (Netting 1993:314-316). Netting admired Chayanov's ideas and found many of them useful, but he seemed to remain uneasy with Chayanov's failure to spell out how the world outside of the household affected farm production and the conditions under which demography mattered, if at all (Netting 1993:318).

When Does Demography Matter?

Netting's concluding meditation on Chayanov is the starting point for the empirical analysis of this chapter. Building on the theoretical and empirical work of economist Dwayne Benjamin (1992), this chapter estimates the extent to which market exposure causes demographic variables to wane in determining farm output and, in so doing, links the conditions under which demography might influence production in an explicit way.

The intuition is simple. When rural households live far from towns, with access to ample land (as did Chayanov's peasants and as do most tropical lowland Amerindians), they face poorly developed markets for credit and labor. Under such conditions, production and consumption overlap—households produce what they consume and consume what they produce. Also under such conditions, the demographic make-up of the household dictates production. Assuming households cannot draw on reciprocity to recruit laborers from outside the family, households are stuck with their own laborers to produce what they need. As Netting (1993) and Chibnik (1987:77,

99) pointed out, Chayanov's idea about the dominant role of demography in production was inspired by (and was meant to apply to) relatively autarkic rural households.

But as markets for labor and credit develop, households have other ways of intensifying production besides relying on their own workers. Households can borrow or hire workers to overcome the constraints imposed by the size and composition of their families. If household and hired labor substitute well for each other (Deaton 1997:183-184), households are no longer constrained by their size or composition. As market economies envelop rural households, household consumption and household production start to diverge and demographics become less relevant in explaining farm output or the allocation of labor. As Benjamin puts it, "With separation (of production and consumption), the number of workers in Baron Rothschild's vineyards should not depend on the number of daughters he has." (Benjamin 1992:288).

Even in modern economies, demography may still affect production (Benjamin 1992:290). If labor and credit markets work poorly, or if household and hired labor do not substitute well for each other because hired labor is too costly to monitor, households may continue using their own workers (López 1986; Pitt and Rosenzweig 1986). If, for these or other reasons, households in modern economies prefer to continue using their own laborers, the size and the composition of the household will affect output. The degree to which household demographics affects production is, therefore, a query that can be settled best through empirical analysis (Deaton 1997:184).

Goals and Econometric Approach

The empirical analysis below tries to accomplish four goals. First, it tries to answer Netting's lingering doubts about Chayanov's model: does demography hold its own after controlling for omitted variables unrelated to the household? This concern is addressed by carrying out two different estimations of how demography affects household production. One estimation includes the distance from the village to the nearest market town, and the other estimation excludes the distance variable. The second goal consists of estimating and comparing the weight of demographic variables between autarkic and non-autarkic households in the pooled sample, to test whether demography gains prominence with weak or missing markets. Third, the analysis estimates and compares the strength of demographic variables in autarkic and non-autarkic settings for each of the five ethnic groups separately, to ensure results from the pooled sample hold up across different cultures. Last, different definitions of demography and of autarky are used to ensure robustness in the empirical results.

To estimate the intensity of production (Y) of household i, a Cobb-Douglas production function is used,

$$Y_i = \beta_1 X_{2i}^{\beta 2} X_{3i}^{\beta 3} e^{ui} \qquad \text{(EQ 7-1)}$$

where

Y	= output on imputed farm income per adult from the principal annual crops
X_2	= labor input (proxied by various type of demographic variables)
X_3	= capital input (proxied by wealth or by the value of physical assets)
u	= stochastic disturbance term
e	= base of natural logarithm

Logarithms were taken on both sides of the equation and ordinary least squares used to estimate parameters.

Variables and Potential Endogeneity

The identification and definition of the variables used in the analysis and a discussion of potential endogeneity follows. Tables 4-1 and 5-1 contain the definitions and summary statistics of the variables used in the present analysis.

Dependent Variable

The dependent variable is the value of imputed production per adult from the chief annual crops (maize, peanuts, rice). An adult is defined as a person over the age of 16.

Explanatory Variables Besides Demography

Besides demographic variables (to be described shortly), explanatory variables included household wealth, formal schooling, and Spanish fluency of household heads. Dummy variables for ethnic groups were also included in the pooled sample.

Explanatory Variables: Demography and Integration to the Market

The following variables or groups of variables were used to capture the demographic composition of the household:

- total household size
- dependency ratio (children/adults)
- total household size and dependency ratio—each included as a separate explanatory variable
- number of boys, girls, adult men, and adult women—each included as a separate explanatory variable
- each of the four demographic groups listed under (4) plus the dependency ratio, or variable #2

Earnings from wage labor were used to define participation in the market.

Potential Endogeneity

Several explanatory variables—such as wealth and household size—may be endogenous, biasing the estimated coefficients (Benjamin 1992:304-305). If higher farm output per adult allows households to grow larger (Chibnik 1987:78-79), for example, the estimated coefficients for demographic variables will be biased upwards, increasing the likelihood of accepting the hypothesis that demography matters, even if it does not.

This study did not have the instruments to resolve the potential endogeneity of variables such as household size. Since the interest lies in comparing the strength of demographic variables between autarkic and non-autarkic households, endogeneity (even if present), should not affect the results—provided the analytical stress lies in comparison of the two samples rather than the level of the coefficients. Endogeneity could still bias results if it is more marked among autarkic households. Although this study could not correct for endogeneity in a direct way, it used different definitions of demography to ensure robustness in empirical results (Benjamin 1992:305). Some demographic variables (e.g., dependency ratio) are less endogenous than others (e.g., household size).

Results

Tables 7-1 through 7-3 contain the regression results. Tables 7-1a and 7-1b contain the results of the analysis for the pooled sample, to see whether demographic variables matter after controlling for an important non-household variable such as distance from village to town. Tables 7-2 and 7-3 contain the results of the comparison between autarkic and non-autarkic households for the pooled sample (table 7-2) and for each ethnic group (table 7-3) to see whether demography matters more in autarkic settings.

Does Demography Matter After Controlling for Distance from Village to Town?

Pooled Sample

Tables 7-1a and 7-1b contain the regression results for the pooled sample. The regressions for each different definition of demography were run with and without a variable measuring distance from village to town. The results shown in columns 1-2 of table 7-1 suggest that household size had almost no effect on farm output. A doubling of household size reduced the value of farm output per adult by 16-18 percent—depending on the model used—but results were statistically insignificant. An increase in the dependency ratio (columns 3-4) also had a minor effect on farm output. A doubling of the dependency ratio increased farm output by about only four to five percent. When both variables—household size and dependency ratio—were included together (columns 5-6), the magnitudes of the elasticities increased but remained statistically insignificant, as shown by the results of the F test. The results shown in columns 7-8, with separate variables for each of the four demographic groups (girls, boys, women, and men), suggest that adult women and adult men had the greatest impact on farm output even after controlling for distance from village to town. The elasticities of farm output, with respect to the number of adult women or men, ranged from –0.35 (adult women) to –0.44 (adult men) and were statistically significant at the 95 percent confidence level or above. In the most complex models (columns 9-10), with separate variables for boys, girls, women, men, and the dependency ratio, the dependency ratio continued to play a weak economic and statistical role (elasticity −0.009 to 0.10 significant at about the 10 percent confidence level).

TABLE 7-1a *Effects of Demography on Farm Output: Pooled Sample*

Variable	Distance- [1] Coef	Se	Distance+ [2] Coef	Se	Distance- [3] Coef	Se	Distance+ [4] Coef	Se	Distance- [5] Coef	Se
HHsize	-.16	.24	-.18	.24					-.46	.32
Dep ratio					.05	.07	.04	.07	.15	.10
Boys										
Girls										
Men										
Women										
Wealth	$.36^3$	.07	$.38^3$	.07	$.35^3$	.07	$.36^3$	.07	$.37^3$	.07
Education	-.06	.05	-.02	.05	-.06	.05	-.03	.05	-.07	.05
Spanish	-.38	.43	-.19	.44	-.36	.43	-.18	.44	-.40	.43
Mojeño	-.10	.31	-.36	.33	-.14	.31	-.39	.33	-.14	.31
Yuracaré	.10	.41	-.08	.42	.06	.41	-.11	.42	-.01	.41
Tsimane´	$-.57^1$	.35	-.12	.41	$-.53^1$	.35	-.14	.41	$-.64^1$	.35
Tawahka	1.86^3	.41	2.28^3	.46	1.77^3	.41	2.18^3	.46	1.84^3	.41
Distance			$.46^1$	.22			$.45^2$	.22		
Constant	3.08^3	.77	.84	1.35	2.90^3	.70	.69	1.30	3.56^3	.84
Prob>F (%)									27.72	
Obs	688		688		688		688		688	
Adj R2	.06		.07		.06		.07		.07	

NOTE   Regressions are ordinary least squares. For definition and summary statistics of variables, see tables 4-1 and 5-1. Dep ratio=children/adults; adult are people over age of 16. Dependent variable is imputed value of farm output per adult. Dependent variable, wealth, demographic variables, distance, and education in logarithms. Dummies used for Spanish and for ethnic groups; Chiquitano is excluded ethnic group. Prob>F is test of significance of joint demographic variables. Distance+ and distance- refers to regression with (distance+) and without (distance-) variable measuring distance from village to town. 1, 2, and 3 significant at the ≤10%, ≤5%, or ≤1%.

TABLE 7-1b Effects of Demography on Farm Output: Pooled Sample

Variable	Distance+ [6] Coef	Se	Distance- [7] Coef	Se	Distance+ [8] Coef	Se	Distance- [9] Coef	Se	Distance+ [10] Coef	Se
HHsize	.46	.32								
Dep ratio	.14	.10					-.009	.15	-.01	.15
Boys			.10	.08	.10	.08	.10	.12	.01	.12
Girls			-.03	.05	-.03	.05	-.04	.07	.03	.07
Men			$-.40^2$	.16	$-.44^3$	.16	$-.40^2$	.17	$.44^3$	.17
Women			$-.35^2$	.12	$-.36^3$	.12	$-.35^3$	.07	$-.37^3$	.12
Wealth	$.39^3$	.07	$.39^3$	.07	$.41^3$	.07	$.39^3$	.07	$.41^3$	.07
Education	-.03	.05	$-.06^2$	.04	-.02	.05	-.06	.05	-.02	.05
Spanish	-.22	.45	-.54	.43	-.33	.44	-.54	.43	-.33	.44
Mojeño	-.39	.33	-.20	.31	-.51	.33	-.20	.31	$-.51^1$	.33
Yuracaré	-.16	.42	.10	.41	-.33	.42	-.10	.41	-.33	.42
Tsimane′	-.20	.42	$-.86^2$	.36	-.35	.41	$-.87^2$	.36	-.35	.41
Tawahka	2.25^3	.46	1.78^3	.41	2.27^3	.46	1.78^3	.41	2.27^3	.46
Distance	$.45^2$	.22			$.54^3$	.22			.54	.22
Constant	1.36	1.38	2.93^3	.70	·27	1.30	2.93^3	.71	.27	1.30
Prob>F (%)	29.87		.47		.22		1.03		.50	
Obs	688		688		688		688		688	
Adj R2	.07		.08		.09		8.41		9.07	

The results from the pooled sample suggest that of all the demographic variables, the number of adult men and women matter most in shaping output—whether or not one includes distance from town to village. Even though some demographic variables (e.g., children, dependency ratio) had a weak effect on farm output in the most complex models (columns 9-10), demographic variables taken as a group affected output—as shown by the results of the F tests. Contrary to Chayanov's prediction, the dependency ratio had an economically and statistically negligible role in production, after controlling for household size and town-to-village distance (columns 9-10). The dependency ratio was not an important determinant of farm output in any of the regressions (columns 3-6, 9-10). The next part of the analysis examines whether demographic effects are stronger in autarkic households.

Comparison of Autarkic and Non-Autarkic Households: Pooled Sample

Table 7-2 contains the regression results for the pooled sample, broken down by whether or not households worked for wages. In table 7-2, each pair of regressions (autarky or market) was run using a different demographic variable or group of demographic variables. Columns 1-2, for example, contain the results of the same regression using household size as the demographic variable—columns 1 and 2 contain the results of the regression applied to autarkic households (column 1) and non-autarkic households (column 2). Autarky refers to households that did not work for wages.

The results of table 7-2 suggest that demography matters more in autarkic than non-autarkic settings, irrespective of how one defines demography. The coefficients, levels of statistical significance of demographic variables, results of F tests, and adjusted R squares all suggest that demographic variables play a more prominent role among autarkic than among non-autarkic households

In columns 1 and 2, for example, one sees that a doubling of household size reduces output per adult by about 72 percent among autarkic households (p>|t|=1.1%), but a similar increase in household size among non-autarkic households increases output per adult by only about ten percent—and the result is statistically insignificant (p>|t|=77.9%). The most complex models (columns 9-10) show a better fit for autarkic households (adjusted R square=0.14) than for households earning income from wages (adjusted R square=0.07). In those two regressions, demographic variables play a statistically stronger role among autarkic (prob>F=3.66%) than non-autarkic households (prob>F=10.10%).

So far the evaluation of results suggests that demographic variables seem to matter more among autarkic households. But how about Chayanov's idea that the ratio of consumers to adult producers affects the intensity of work? The discussion in the previous section suggested that the dependency ratio is not an important determinant of output once we control for village-to-town distance and different demographic groups within the household (columns 9-10 of table 7-1). The sign, size, and level of statistical significance of the

TABLE 7-2a *Effects of Markets and Demography on Farm Output: Autarkic and Non-Autarkic Households Compared*

Variable	Autarky [1] Coef	Se	Market [2] Coef	Se	Autarky [3] Coef	Se	Market [4] Coef	Se	Autarky [5] Coef	Se
HHsize	-.72³	.28	.10	.37					-1.14³	.38
Dep ratio					-.04	.08	-.11	.12	.19	.11
Boys										
Girls										
Men										
Women										
Wealth	.38³	.08	.34³	.11	.33³	.08	.34³	.10	.41³	.08
Education	-.03	.05	-.03	.07	-.04	.06	-.03	.07	-.04	.05
Spanish	1.31	.48	1.27	.68	1.28³	.49	-1.27¹	.67	1.31³	.48
Prob>F(%)									1.07	
Obs	283		405		283		405		283	
Adj R2	.14		.06		.12		.06		.14	

NOTES   Regressions are ordinary least squares with constant and with dummies for ethnic groups(not shown). For definition and summary statistics of variables, see tables 4-1 and 5-1. Dep ratio=children/adults. Dependent variable is imputed value of farm output per adult. Dependent variable, wealth, demographic variables, and education in logarithms. Spanish is dummy. Prob>F is test of significance of joint demographic variables. Autarky=households that did not work for wages. 1, 2, and 3 significant at $\leq$10%, $\leq$5%, or $\leq$1%.

coefficient for the variable measuring dependency ratio in table 7-2 allows one to explore Chayanov's idea in greater depth.

The coefficients of the variable for the dependency ratio in table 7-2 (columns 3-6 and 9-10) suggest that the dependency ratio continues to have a statistically weak effect on output in both autarkic and non-autarkic households. Among autarkic and non-autarkic households, the elasticity of output with respect to the dependency ratio ranged from –0.04 to 0.19—all results were statistically insignificant at the 90 percent confidence level.

TABLE 7-2b *Effects of Markets and Demography on Farm Output: Autarkic and Non-Autarkic Households Compared*

Variable	Autarky [6] Coef	Se	Market [7] Coef	Se	Autarky [8] Coef	Se	Market [9] Coef	Se	Autarky [10] Coef	Se
HHsize	-.15	.47								
Depratio	.14	.15					.08	.17	.03	.23
Boys			-.09	.10	$.20^1$	.12	-.15	.15	.18	.17
Girls			-.01	.05	-.03	.07	-.04	.08	-.04	.11
Men			$-.38^3$	.15	-.13	.34	$-.37^3$	.15	-.11	.37
Women			$-.26^1$	.14	$-.49^3$	.19	$-.24^1$	.14	$-.48^1$	.19
Wealth	$.34^3$	.11	$.38^3$	.08	$.36^3$	.11	$.39^3$	.08	$.36^3$	.11
Education	-.04	.07	-.05	.05	-.02	.07	-.05	.05	-.02	.07
Spanish	-1.30^1	.68	1.22^3	.49	-1.49^2	.68	1.22^3	.49	-1.48	.68
Prob>F(%)	62.48		2.0		5.63		3.66		10.10	
Obs	405		283		405		283		405	
Adj R2	.06		.15		.07		.14		.07	

Comparison of Autarkic and Non-Autarkic Households: Results by Ethnic Group

Since the results for the pooled sample discussed so far may gloss over differences between ethnic groups, the study must analyze whether the results from the previous sections about the stronger role of demography in autarky also hold in each ethnic group. Table 7-3 contains a summary of the regression results for each ethnic group, broken down by whether or not households worked for wages. The regression models used in table 7-3 are the same as the regression models used in table 7-2, except that they were run for each separate ethnic group. The more complex models of table 7-2 (e.g., columns 5-10) could not be used because the sample size of some ethnic groups was too small. A parsimonious model (columns 1-4 of table 7-2) was used instead, with only two demographic variables (household size and dependency ratio)—included separately for each ethnic group.

TABLE 7-3a Summary of Effect of Demographic Variables on Farm Output: Results by Ethnic Group

	Obs	Coef	Se	P>|t| (%)	Adj R2
Tawaha					
Autarky					
HHsize	31	-.77	.85	37.5	37.90
Dep ratio	31	.11	.80	88.9	36.00
Market					
HHsize	54	1.75	1.05	10.3	4.76
Dep ratio	54	1.82	.45	0.1	24.43
Tsimané					
Autarky					
HHsize	71	.48	.38	20.3	39.45
Dep ratio	71	-.009	.11	93.5	37.94
Market					
HHsize	100	-.94	.79	24.0	3.31
Dep ratio	100	.008	.25	97.2	1.88
Mojeño					
Autarky					
HHsize	86	-.54	.67	41.8	-1.22
Dep ratio	86	.06	.19	73.9	-1.91
Market					
HHsize	46	.29	1.52	84.6	-.35
Dep ratio	46	.12	.44	78.4	-.26

NOTES Models 1-4 from table 7-2 used. For definition and summary statistics of variable see tables 4-1 and 5-1. Same notes as in table 7-2, except no dummies included for ethnic groups. Table only contains coefficients of demographic variables; coefficients for all other variables not shown.

According to the results of tables 7-3a and 7-3b, no ethnic group supports the idea that demography is more significant in autarky or that the dependency ratio drives production. Except for the Tawahkans with the closest link to the market, demographic variables did not seem to affect farm output for either autarkic or non-autarkic households. The elasticities of output with respect to demographic variables for all the groups in Bolivia were generally small and, in all cases, statistically insignificant. Among the Tawahka, demographic variables mattered more among households with links to the market. Among Tawahkan households with links to the market, a doubling of household size or the dependency ratio increased output per adult by

TABLE 7-3b *Summary of Effect of Demographic Variables on Farm Output: Results by Ethnic Group*

	Obs	Coef	Se	P>ltl (%)	Adj R2
Chiquitano					
Autarky					
HHsize	60	-.72	.67	29.1	-2.03
Dep ratio	60	-.19	.17	27.6	-1.90
Market					
HHsize	178	.14	.55	79.2	8.69
Dep ratio	178	.04	.18	79.3	8.69
Yuracaré					
Autarky					
HHsize	35	-.63	1.00	53.2	-3.21
Dep ratio	35	-.23	.49	63.8	-3.80
Market					
HHsize	27	-.86	1.36	53.2	7.74
Dep ratio	27	-.35	.39	38.2	9.26

about 175 to 180 percent (p>ltl=0.1% for dependency ratio and 10.3% for household size).

Conclusion

It is time to return to the two concerns Netting raised in the conclusion of his last book and see if either of them have been resolved.

1. Does demography matter after controlling for confounding covariates, particularly for variables unrelated to the household?

The evidence from the pooled sample (tables 7-1a and 7-1b) suggests that demography matters even after controlling for wealth, human-capital attributes of household heads, ethnicity, and at least one community variable (distance from village to town). Among demographic variables, the number of adult women and men mattered more than the number of children or the ratio of consumers to adult producers. Chayanov's idea about the role of the dependency ratio was correct. As the dependency ratio grows, so generally does the farm output produced by each adult worker. Although the variable dependency ratio bore the correct, positive sign predicted by Chayanov in most of the models of tables 7-1a and 7-1b, the coefficient was physically and statistically insignificant.

2. Are Chayanov's ideas cross-culturally robust, both in autarkic and non-autarkic settings?

The evidence from the pooled sample presented in tables 7-2a and 7-2b suggest that demographic variables seem to be more significant in autarkic households. The dependency ratio, however, does not seem to have a significant effect on the output of adult workers among autarkic or non-autarkic households in either the pooled sample or any ethnic group.

The results echo the findings of Chibnik (1987:89). Using information from about a dozen societies and bivariate analysis, Chibnik found that the correlation between the ratio of consumers to adult workers and output per worker was statistically significant at the 95 percent confidence level or above for only five societies. In at least one case, the sign of the coefficient for the dependency ratio was negative.

The weaker statistical results for separate ethnic groups may stem from the small sample size of each ethnic group and perhaps from poor measurement of capital and labor inputs—faults that may become more visible in smaller samples. Wealth and demographic composition was used to measure capital and labor inputs. Better measures would have been the actual capital and labor inputs used to produce crops (Durrenberger 1979; Minge-Kalman 1977; Tannenbaum 1984).

The next chapter tackles this shortcoming, but from another angle. If the ratio of consumers to adult producers affects the intensity of work—as Chayanov said—then the ratio should also affect leisure, which is the complement of work. More dependents per adult worker should cause leisure to decline. The next chapter tests the idea, using panel information from spot observations of the Tawahka.

This chapter brings to a close the study of how markets affect the way indigenous people use natural resources, and how markets may dampen the role of demography in production. Chapters 5-7 have tried to show that markets produce predictable outcomes in some areas. Even among relatively isolated indigenous people in the tropical lowlands of Latin America, markets seem to produce a Kuznets curve of deforestation (see chapter 5, "Forest Clearance: Income, Technology, and Private Time Preference"). Markets also seem to weaken the role of demography in production in ways that were anticipated but not fully articulated or tested by Chayanov or Netting (discussed in this chapter). This analysis has also shown that economic development will continue to exert pressure on most types of edible game in the forest, although the pressure will probably weaken as incomes rise (see chapter 6, "Game Consumption, Income, and Prices: Empirical Estimates and Implications for Conservation").

With the section on infrastructure closed, it is time to explore how markets affect aspects of social life. To do this, we re-open Marshall Sahlins' hypothesis about the original affluent society and—using a multivariate approach on a panel data set on scans—test whether leisure declines with modernization.

Chapter 8

Chayanov and Sahlins on Work and Leisure

When comparing cultures with different types of subsistence or economies with different levels of income, the amount of leisure available to people provides an intuitive feeling for whether the quality of life gets better or worse with economic development and cultural change. In cultural anthropology, Marshall Sahlins immortalized the idea of using leisure as a yardstick to gauge prosperity and the quality of life. In his justly famous, provocative, and terse masterpiece, "Notes on The Original Affluent Society," Sahlins (1968) reviewed the archaeological and ethnographic record and concluded that leisure declined with economic complexity—as one went from foraging bands to swidden systems and on to industrial societies. Sahlins put it this way:

> Extrapolating from ethnography to prehistory, one may say as much for the Neolithic as John Stuart Mill said of all labor saving devices: that never was one invented that saved anyone a minute's labor. The Neolithic saw no particular improvement over the Paleolithic in the amount of time required per capita for the production of subsistence; probably, with the advent of agriculture people had to work harder.
>
> (Sahlins 1972:35 quoted in Sackett 1996:36)

Like Chayanov's household developmental cycle, Sahlins's affluent foragers soon became a stylized fact in anthropology (Sackett 1996:37-38).

Sahlins's idea that modernization erodes leisure deserves scrutiny because the putative link between the complexity of subsistence and leisure may shrink once one controls for the role of variables that affect subsistence and leisure. One such omitted variable may be Chayanov's ratio of consumers to producers. Chayanov did not analyze leisure, but he did analyze the other side of leisure—or the intensity of work. Chayanov said that the intensity of work reflected household demographics, with leisure decreasing as the number of dependents on each adult worker rose (see chapter 7, "Chayanov and Netting: When Does Demography Matter?"). According to Sahlins, the amount of leisure reflected the type of subsistence or the degree of economic development, with leisure decreasing as one moved up the scale of economic complexity.

The two views are, of course, compatible. The type of subsistence and household demographics could both affect the amount of leisure. As discussed in the last chapter, demographics may matter, but chiefly with missing or with imperfect markets. A robust test of Chayanov and Sahlins's hypotheses requires estimating the many determinants of leisure or work and comparing the degree to which variables related to household demographics (Chayanov) or to the complexity of the economy (Sahlins) affect the outcome. Anthropologists have yet to do such a test.

The failure to do so probably reflects the difficulties of combining information from traditional anthropological methods used to study time allocation (e.g., scans) and consumption (e.g., weigh days) with information from surveys on demography or human capital. The creation of a primary information file on time allocation suitable for multivariate analysis requires merging, collapsing, aggregating, and expanding uneven information of different types from different sources. Spot observations, for example, come from blocks of time chosen at random several times a month. Information about demography and wealth come from periodic surveys. Information on consumption and income may come from weigh days (consumption) or from surveys done every week or month (cash income). Researchers may take readings of temperature and rainfall several times on the same day. Combining information of such different forms into one computable database was hard until the advent of modern computer software.

Cross-Cultural Evidence and Theory

Sahlins's idea that the amount of leisure declines with economic development does not fit well with the best and most recent information. A recent review by Ross Sackett (1996) of 102 anthropological studies of personal time allocation in 76 societies provides only partial backing for Sahlins. Sackett shows that leisure declines as one moves from foraging bands to societies practicing swidden agriculture (as Sahlins said), but then increases in industrial societies.

Evidence from industrial nations lends further support to Sackett's finding that economic complexity may be associated with more leisure. Drawing on household surveys in Japan, the United States, Norway, Denmark, and the former Soviet Union, Juster and Stafford (1991) showed that the amount of leisure increased by large amounts for men and women from the 1960s to the 1980s. Except for men in the Soviet Union (for whom leisure declined), the amount of leisure increased by about 10 percent for men and 16 percent for women in all countries. The finding of Juster and Stafford fits with trends since the beginning of the twentieth century in the United States documented by economist Dora Costa (1997, 1998). Costa showed that the real price of leisure has declined over the past one hundred years, allowing people—particularly the poor—to enjoy more leisure.

The empirical evidence seems to show that leisure need not decline with economic development. Theoretical reasoning should also lead one to question Sahlins's idea. Economic development may increase or decrease leisure because of income and substitution effects that work in opposite directions (Becker 1965). As economies develop and incomes rise, people want more leisure because leisure is a normal good (income effect). But higher income also raises the value of people's time, causing people to curtail their leisure (substitution effect). The net effect of economic development or income growth on leisure will depend on the relative strength of the income and substitution effects. If the income effect outweighs the substitution effect, people with higher income or those living in more complex economies will enjoy more leisure. If the substitution effect overshadows the income effect, however, people will enjoy less leisure.

Goals

The rest of this chapter tests the hypotheses of Chayanov and Sahlins by estimating the simultaneous effects of 1) economic development (Sahlins) and 2) the ratio of consumers to adult producers (Chayanov) on the amount of time indigenous people devoted to farm work and leisure. Panel information from the Tawahka was used to achieve these goals.

Since work and leisure are two sides of the same coin, one could have tested the hypotheses by analyzing the determinants of only one of the two activities. This study chose both activities to deal with errors in the measurement of the dependent variable (work), to remain faithful to the spirit of Chayanov and of Sahlins, and to ensure robustness in the empirical results.

The dependent variable for work contains more measurement errors than the dependent variables for leisure, and should be read with care. The two types of variables came from scans, as described in chapter 3, "Research Design." Scans produce more reliable information when researchers can see the activities they jot down. When subjects leave the village to hunt or farm, researchers doing scans must infer what people were doing from third parties because they cannot see the subject. Scans produce less accurate information when used to measure the amount of time people devote to activities—such as farm work—that take place outside of the village or that researchers cannot observe in a direct way. Since farm work takes place outside of the village, most of the information collected on farm work (85%) came from third parties. On the other hand, leisure generally takes place in the village, so researchers using scans are more likely to see subjects resting. The information will therefore contain less measurement error. Although information on the amount of time devoted to leisure and farm work both came from scans, information on leisure came mainly from direct observations (and is therefore more reliable), whereas most of the information on farm work came from third parties.

Leisure and farm work were also examined as two separate, dependent variables to remain faithful to the ideas of Chayanov and Sahlins. Chayanov wrote about the effect that the ratio of consumers to adult workers had on the intensity of farm work. Sahlins, on the other hand, wrote about the effect of economic development or cultural evolution on leisure.

Econometric Approach

To test the hypotheses of Sahlins and Chayanov, four separate multivariate binomial logit regressions were run, each with a different dichotomous dependent variable. The four dependent variables included three different types of leisure (to be described shortly) and farm work. Since the information came from repeated observations of the same sample of people over two years, a random-effect model was used with an equal-correlation or exchangeable structure, and with robust standard errors adjusted for clustering by subject.

Information and Variables

All four dependent variables (three for leisure and one for work) came from scans. Explanatory variables came from different sources. The study drew on demographic surveys to construct the ratio of consumers to adult workers, estimate household size, and create variables that would capture the attributes of the person and the household. The study also drew on the time and date when researchers carried out the scans (or saw the subjects) and on daily readings of temperature and rainfall in each village. Climatological and time variables were used because temperature, rainfall, season, and time of day have obvious effects on leisure and farm work. Last, the study drew on monthly surveys of personal income to create a variable for cash income, which was lagged by one month to reduce potential endogeneity.

Dependent Variables

The variable work took the value of one if the subject was doing a farm chore when the researcher saw the subject, and zero otherwise. Information on farm work came from direct and indirect observations. As mentioned, 85 percent of the information on farm work came from third parties and should be considered less reliable than information on leisure.

Like farm work, leisure was also measured as a dichotomous variable. Since one could define leisure in many ways, the analysis was done using three different definitions to ensure consistency in results. First, leisure was equated with social and personal activities unrelated to work but necessary for social reproduction—such as visiting, playing, socializing, eating, and grooming. In the tables and discussion below these activities are called *leisure1*. Second, leisure was equated with being inactive, static, idle, or resting; I called these activities *idle*. Last, a third category was created—*leisure2*—which combined the first two categories, *leisure1* and *idle*. In the analysis of leisure, only information from direct observations is used.

Explanatory Variables

A variable was created to test Chayanov's hypothesis for the ratio of dependents to adult producers in the household. To identify a dependent, the amount of time different age cohorts spent working on farm chores was estimated. Although Tawahka children as young as five years of age already spend 3 percent of daylight time (6am-6pm) on farm work. The amount of time children spent on farm work rose by the age of nine. Five to eight-year old children spent an average of 4.14 percent of their time working on farm chores, but children nine to 15 years of age spent more than twice as much time working (10.25%). Children under eight years of age were classified as full dependents or net consumers, even though they did some farm work. Since Chayanov's theory about the increasing drudgery of work refers to adults, the statistical analysis of work and leisure was limited to people over the age of 15—roughly the age at which the Tawahka become adults.

Testing Sahlins's hypothesis is more difficult if one wants to remain faithful to the spirit of his idea. When Sahlins noted in the earlier quote that leisure declined with modernization, he was referring to changes taking place in a grand scale of time—from the Paleolithic to the present. Since the ethnographic information used in this study comes from the present, one must use different indexes of economic modernization as explanatory variables to capture the spirit of the process Sahlins was trying to underscore.

Several variables or groups of variables were used to proxy for economic development. First, the joint effect that several modern forms of human capital (e.g. formal schooling, competence in written and oral Spanish, knowledge of arithmetic) had on leisure or work was estimated. Second, the effect of last month's cash income on the present likelihood of doing farm work or being found in leisure activities was estimated. Third, the study examined the coefficient of the dummy variable for the village of Yapuwás—the poorer of the two Tawahka villages. If leisure is a dependent variable and Sahlins is right, then the coefficients of cash income and modern forms of human capital should have negative signs in the logit regression, and the coefficient on the dummy variable for the more traditional village of Yapuwás should have a positive sign.

Other explanatory variables included demographic attributes of the person, household size, total daily rainfall, mean daily temperature, and dummy variables for the time of the day, day of the week, seasons of the year, and the villages. Weather variables were interacted with dummy variables for seasons. Table 8-1 contains definition and summary statistics of the variables used in the logit regression.

Potential Endogeneity and Fixed Effects

The estimates may contain two biases: endogeneity and failure to control for unobserved personal fixed effects. Most of the explanatory variables are exogenous because they lie outside of the subject's control. That is not so with income, household size, or even with the dependency ratio. One's income affects how much leisure or work one has at present, but one's level of past work or leisure will affect one's present income. To resolve the potential bias, values of income lagged by one month were used as predictors of today's level of work or leisure. Though the approach does not get rid of the bias if past and present levels of income, work, and leisure and their error terms interact over time, lagging partially mitigates the bias.

Another bias may arise from failure to control for a subject's unseen, fixed characteristics. If the unseen propensity for idleness is positively related to leisure and negatively related to income, failure to control for the propensity will bias the estimated coefficient on income by lowering it. To correct for the potential bias, fixed-effect logit models were used. The fixed-effect model produced a coefficient for income similar to the one produced by the random-effect model. In the fixed-effect logit estimation, each subject is assigned a dummy variable, which picks up all the unseen, fixed attributes of the subject. The results of the fixed-effect estimation are not reported because many of the variables needed to test the hypotheses could not be estimated with a fixed-effect model. With the information from the Tawahka, the fixed-effect model is able to estimate variables such as income. It is inadequate for estimating the effect of variables such as literacy, numeracy, fluency in Spanish, or the ratio of consumers to producers because those variables did not change (or did not change much) during the two years of observation.

TABLE 8-1 *Definition and Measurement of Variables for Tawahka Over the Age of 15*

Variable	Definition	Obs	Mean	Sd
Idle	Static, resting	8389	.31	.46
Leisure1	Playing, eating, socializing, grooming, visiting	8389	.15	.35
Leisure2	Idle+leisure1	8389	.46	.49
Work	Farm work	15708	.09	.29
Age	Age in years	110	29	13
Male	Sex of subject	111	.50	.50
Education	Maximum schooling	106	3.24	2.66
Numerate	Skills in arithmetic	107	.71	.45
Fluespan	Fluency in Spanish	108	.70	.45
Spanlit	Literate in Spanish	108	.46	.50
Income	*lempiras*/month/person	500	2883	5888
Dep ratio	Ratio of dependents to adult; see text	32	.98	.51
HHsize	Household size	32	9.1	4.2
Monday	% of scans on Monday	8389	.18	.39
Tuesday	% of scans on Tuesday	8389	.13	.34
Wednesday	% of scans on Wednesday	8389	.13	.33
Thursday	% of scans on Thursday	8389	.12	.32
Friday	% of scans on Friday	8389	.19	.39
Saturday	% of scansd on Saturday	8389	.12	.32
Sunday	% of scans on Sunday	8389	.10	.31
Afternoon	% observations in afternoon (2-6pm)	8388	.15	.36
morning	% observations in morning (6-10am)	8388	.47	.49
Midday	% observations in midday (10am-2pm)	8388	.36	.48
Rain mm	Daily rainfall in mm	1228	6.17	10.14
Avg temp	Average of minimum and maximum daily temperature in centigrades	857	26.8	1.93

NOTES   Following dummy variables (name of dummy variable = 1) come from scans or spot observations: idle, leisure1, leisure2, work, day of week, time of day. The following dummy variables come from surveys: numerate, fluespan, spanlit, sex(male), income.

Results

Table 8-2 contains the regression results. The discussion is split into two parts—one dealing with Chayanov's hypothesis and the other dealing with Sahlins' hypothesis.

TABLE 8-2 *Determinants of Leisure and Work: Results of Random-effect Logit Regressions, Tawahka*

Dep Variable	Work [1]		Leisure1 [2]		Leisure2 [3]		Idle [4]	
Variable	Coef	Se	Coef	Se	Coef	Se	Coef	Se
Dep ratio	-.24	.21	-.46	.11[3]	-.32	.10[3]	-.12	11
Age	-.17	.02[3]	-.04	.01[3]	-.01	.01	-.05	.01[3]
Age2	-.06	.02[3]	-.0005	.0002[3]	.0002	.001	.0007	.0009[3]
HHsize	-.06	.02[3]	.002	.01	.01	.01	-.001	.01
Income	-.03	.01[3]	-.0009	.01	.01	.01	.009	.01
Male	1.29	.25[3]	.45	.11[3]	1.18	.11[3]	1.00	.13[3]
Yapuwas	-1.52	.26[3]	.34	.15[2]	.27	.13[2]	-.19	.14
Education	-.10	.05[1]	.01	.02	.01	.03	-.01	.02
Numerate	-.34	.37	.29	.16[1]	.14	.13	-.003	.15
Fluespan	.08	.25	.35	.16[2]	-.04	.14	-.23	.15
Spanlit	-.07	.31	-.01	.17	.12	.14	.04	.16
Monday	2.12	.16[3]	-.91	.14[3]	-.65	.10[3]	-.15	.12
Tuesday	1.95	.17[3]	-1.16	.13[3]	-.67	.12[3]	.008	.13
Wednesday	1.72	.17[3]	-1.00	.15[3]	-.59	.11[3]	-.01	.12
Thursday	1.79	.15[3]	-.94	.13[3]	-.75	.11[3]	-.24	.11[2]
Friday	1.86	.19[3]	-1.00	.15[3]	-.50	.10[3]	.07	.13
Saturday	1.44	.18[3]	-1.12	.18[3]	-.70	.13[3]	-.05	.13
Morning	1.22	.11[3]	-.42	.11[3]	-.41	.10[3]	-.27	.09[3]
Midday	.38	.12[3]	-.37	.12[3]	-.05	.11	.10	.10
Wetrain	.004	.003	-.004	.003	.007	.003[2]	.008	.003[3]
Wetemp	.08	.03[3]	-.03	.03	-.005	.02	.01	.01
Wet	-2.52	.88[3]	.85	.83	.22	.59	-.43	.57
Obs	8409		4953		4953		4953	
Wald(Prob>chi2)	3.01%		.22%		21.77%		21.83%	

NOTES Regressions are random-effect logits with exchangeable correlation structure, no constant, and standard errors adjusted for clustering by subjects. Income is expressed in natural logarithms. Wald is test for joint significance of numeracy, education, and fluency and literacy in Spanish. 1, 2, and 3, significant at ≤10%, ≤5%, and ≤1%.

Chayanov

To test Chayanov's hypothesis about the dominant role of the ratio of consumers to adult workers on the intensity of farm work, the study focuses on column 1 of table 8-2. The result suggests that the ratio of consumers to workers lowers rather than raises the likelihood of finding an adult working during a scan (coefficient -0.24). Besides having the wrong sign, the estimated coefficient was statistically insignificant ($p>|z|=25\%$). Model 1 was re-run, using the more reliable variable for work based only on direct observations. In that model, the sign of the coefficient for the dependency ratio became positive (0.13), but remained statistically insignificant ($p>|z|=60.5\%$).

If one uses leisure instead of work as a dependent variable, one finds stronger confirmation of Chayanov's hypothesis. The coefficients for the dependency ratio (using *leisure1* and *leisure2* as dependent variables) had the correct negative signs and were statistically significant at the 99 percent confidence level. Adults with more dependents were less likely to be found in leisure activity—as Chayanov might have predicted. For the type of leisure called *idle*, the variable for dependency ratio had the correct negative sign but was statistically insignificant (coefficient -0.12; $p>|z|=27.9\%$).

In summary, the information on time allocation does not allow acceptance of Chayanov's hypothesis that the ratio of consumers to adult workers affects the likelihood of doing farm work. The dependency ratio, however, seems to affect leisure in the way Chayanov might have predicted. Adults with more dependents were less likely to be found resting.

Sahlins

Sahlins' idea that greater participation in a complex economy reduces leisure was tested by estimating the effect on leisure of: 1) modern forms of human capital, 2) cash income, and 3) living in the poorer and more isolated village of Yapuwás.

First the variables related to modern forms of human capital were examined. The model using *leisure1* as a dependent variable shows that only knowledge of arithmetic and fluency in Spanish were statistically significant predictors of leisure, but they had a different sign than Sahlins might have predicted. Knowledge of arithmetic (coefficient 0.29; $p>|z|=7.1\%$) and fluency in Spanish (coefficient 0.35; $p>|z|=2.6\%$) were associated with a greater likelihood of being found in leisure. In the model using *leisure2* as a dependent variable, none of the human-capital variables were statistically significant predictors of leisure at the 90 percent confidence level or above, and three of the four variables had the wrong (positive) sign. In the model using *idleness* as a dependent variable, all the human-capital variables except for literacy had the right (negative) sign, but none were statistically significant. The joint statistical significance of all human capital variables was tested for, and the null hypothesis of no effect on leisure for two of the three models was accepted. Only when using *leisure1* as a dependent variable did modern human capital seem to affect leisure (prob>chi2=0.22%).

Next, the effect of last month's cash income on each of the three forms of leisure was examined and no support for Sahlins's hypothesis was found. Lagged cash income only had the correct (negative) sign in the first model, but the coefficient was trivial (–0.009) and statistically insignificant (p>|z|=94.3%). When using *idle* and *leisure2*, the sign of the coefficients for lagged cash income was positive and statistically insignificant (coefficient 0.01, p>|z|=15.8% for *leisure2* and coefficient 0.009, p>|z|=49.1% for *idle*).

Last, the sign of the coefficient for the dummy variable for the village of Yapuwás was examined. The dummy variable had the correct (positive) sign in the models using *leisure1* and *leisure2*, and was statistically significant at the 95 percent confidence level or above. Only when using *idle* as a dependent variable did the coefficient for the dummy variable for Yapuwás have the wrong (negative) sign and was statistically insignificant (coefficient –0.19; p>|z|=18.8%).

In summary, none of the other more direct measures of economic modernization (except for the village dummy) had the effects on leisure one might have expected from Sahlins' hypothesis. Irrespective of how one defines leisure or economic modernization, the two variables seem to be linked in a tenuous way.

Conclusion

The results of the analysis seem to lend greater credence to Chayanov's than Sahlins' hypothesis. As Chayanov might have predicted, the ratio of consumers to adult producers affects the amount of leisure—however one defines leisure. Contrary to Sahlins' expectation, different indexes of economic prosperity had a negligible or wrong effect on leisure or farm work. The finding that economic modernization has unclear effects on leisure or work is unsurprising since price theory and the empirical studies reviewed earlier had already uncovered the ambiguity.

At least two lessons emerged from the analysis. First, anthropologists who have studied the determinants of leisure have been understandably lured by the elegance and simplicity of Sahlins' and Chayanov's ideas. In exploiting these lodes, anthropologists have left alone more pedestrian (but potentially rich) leads. For example, consider the strong role that prosaic variables—such as the time of day or the day of the week—have on work or on leisure (see table 8-2). Among the Tawahka, much leisure takes place in the afternoons and on Sundays. Those variables swamp the variables associated with economic prosperity in determining leisure.

Consider the life-cycle dimensions of leisure. In most of the regressions in table 8-2, age bore a strong, statistically significant, U-shaped relation to leisure. As is true among people in industrial societies (Ghez and Becker 1975; Hill 1985a, 1985b), the Tawahka seem to work hardest in mid-life—when earning potential and physical strength peak. Mid-life may be the time when the Tawahka work hardest to invest in assets for use later in life. They do not seem to be smoothing leisure over their life cycle.

Second, consider the quality rather than the mere quantity of leisure. Future research may prove Sahlins' theory that the amount of leisure declines with economic development, but a decline in the quantity of leisure could be offset by a rise in the quality of leisure—producing unclear effects on the quality of life. An idle day in jail contains many hours of pure leisure of little value.

Since the information presented does not seem to show that economic development erodes the amount of leisure available to indigenous people, this study turns now to an analysis of how economic development might affect health—a less ambiguous index of well being than leisure.

Chapter 9

Human Health:
Does It Worsen with Markets?

For more than half a century, cultural and biological anthropologists and medical professionals have been debating the effects of markets and culture change on the health of indigenous people (Lambert 1931 quoted in Bodley 1988). In recent years, the debate has broadened to include a discussion of how economic development embodied in deforestation, urbanization, and migration contributes to the emergence and re-emergence of epidemics (Brinkman 1994; Levins 1994; Institute of Medicine 1997)—particularly among indigenous people with tenuous links to the market (Jenkins 1989). Through mechanisms we do not understand well, economic development seems linked to the spread of vector-borne diseases (Walsh, Molyneux, and Birley 1993).

The debate about how markets affect health merits attention because it reflects an older, broader, and at times bitter debate in anthropology about the role of ecology, economics, social organization, and ideas in shaping behavior (Harris 1997; Minnegal 1996; O'Meara 1997). This chapter presents a method for estimating and comparing the separate and simultaneous effect that markets and acculturation have on health. To help advance the debate beyond its present stalemate, information is used from 966 household heads of the Tsimane´, Mojeño, and Yuracaré from the river Sécure and the Chiquitano—all in the Bolivian lowlands.

The Three Positions in the Debate

Three positions have emerged in the debate about how markets and acculturation affect the health of indigenous people. Some researchers have said that acculturation and markets undermine the health of indigenous people (Bodley 1988; Confalonieri, Ferreira, and Araújo 1991; Kroeger and Barbira-Freedman 1982; Wirsing 1985). Over a decade ago, Rebecca Holmes (1985:239) summarized the reasons why the health of indigenous people might deteriorate with exposure to markets and acculturation. She noted that modernization led to a sedentary life and produced degenerative diseases, such as obesity, diabetes, and hypertension—all of which were less prominent in traditional societies (Santos and Coimbra 1996:855). Second, breast-feeding declined with modernization and exposed infants to nutritional stress. Third, traditional populations controlled fertility through cultural norms—such as prolonged breast-feeding—that reduced population pressure. Last, contact with Western cultures exposed indigenous people to new diseases for which they lacked immunity. The changes went hand-in-hand with the loss of land and natural resources, the breakdown in the network of social support (Bruhn and Wolf 1979), and a reduction in the biological and ecological complexity of the environment.

About the same time that Holmes published her findings, Rodolf Wirsing reviewed the anthropological, epidemiological, and medical literature on the health of traditional societies undergoing acculturation, and concluded that acculturation worsened health. He noted that:

> Many cultural changes have been detrimental to health and nutrition. Among these are the adoption of a sedentary life-style, the growing preference for imported carbohydrate staples, the increasing use of infant formulas, the reluctance to breast-feed for long periods, the abolition of native family-planning methods, the acceptance of paid work, and the eagerness to grow cash crops.
>
> (Wirsing 1985:315)

A decade later, Shephard and Rode (1996) echoed the findings of Wirsing. In a study of the effects of modernization on circumpolar people, they acknowledged that acculturation improved nutrition (1996:61, 104-105) but noted that circumpolar people enjoyed better health before contact:

> The nature of the arctic habitat and the lifestyle of traditional circumpolar populations protect them against cer-

tain chronic medical disorders that are common in the urban environment. However, contact with the bacteria and viruses of developed societies has brought devastating epidemics to populations that previously lacked immunity to the diseases in question. Currently, acculturation to certain adverse habits of 'modern' society (over-eating, lack of physical activity and cigarette smoking) and adoption of a poorly chosen 'market' diet are also increasing the prevalence of such chronic diseases as ischemic heart disease, diabetes, and several types of cancer....

(Shephard and Rode 1996:249-250)

It is not clear—from the above quote or from their book—whether the supposedly poorer health of circumpolar people who modernize should be blamed on the adoption of "certain adverse habits" of the modern world, an increase in cash income, a sedentary life, or a combination of all three causes. Like Wirsing before them, Shephard and Rode did not disentangle and estimate the separate and simultaneous effects of markets and acculturation on health. They used the two concepts as synonyms.

A second group of researchers—typically development economists—has said that greater participation in a market economy improves health because it raises income and education. This allows households to buy modern medical services and to make better use of inputs to improve health, nutrition, and hygiene (Berry et al. 1987; Santos and Coimbra 1991; von Braun and Kennedy 1994). Some have tempered the position by saying that the magnitude of the improvement may be smaller than the empirical estimates suggest (Strauss and Thomas 1988). By raising income, economic development also increases the demand and price of medical services and medicines, making it harder for the poor to improve their health (Behrman and Deolalikar 1987).

Support for the second position comes from researchers who have shown that acculturation improves nutrition—measured by body-mass index (kg/mt^2) (Baker, Hanna, and Baker 1986:270-271; Norgan 1994; Shephard and Rode 1996:104-105). An international comparison of 10,014 adults in Nigeria, Cameroon, Jamaica, St. Lucia, Barbados, and the United States showed that the body-mass index of men and women over 25 years of age rose as one moved from countries with low income and poor education to countries with high income and better education (Cooper et al. 1997:162).

A third group of researchers has said that the effect of acculturation and markets on the health of rural people is ambiguous, may change over time, and may vary depending on the degree and type of a person's integration to the market (Brown and Whitaker 1994; Fleming-Morán, Santos, and Coimbra 1991; Leatherman 1994; Leatherman, Carey, and Thomas 1995; Packard and Brown 1997; Santos and Coimbra 1996).

Eileen Kennedy (1994) reviewed information from poor countries and concluded that the commercialization of agriculture did not increase child morbidity—the relation varied between countries. In Italy, Brown and Whitaker (1994) showed through historical research that morbidity from malaria and pellagra was not correlated in a linear way with changes in agricultural technology. Morbidity increased but then decreased as the agricultural transformation unfolded. In the southern Peruvian Andes, Leatherman and his colleagues showed (through a cross-sectional study) that exposure to markets produced unclear effects on the nutrition and health of Aymara Indians (Leatherman, Carey, and Thomas 1995). The health of richer villagers improved more than the health of poorer villagers.

Among the Suruí Indians of southwestern Amazonia, greater participation in the market improved nutrition—but only for the more affluent villagers (Santos and Coimbra 1996). The adoption of sedentary life did not worsen nutrition for the northern Aché foragers of eastern Paraguay. Because they had lower nutritional requirements, farming households consumed less meat than foraging households but did not show signs of food deprivation (Hawkes et al. 1987:156). Holmes's (1985) study (mentioned earlier) showed that modernization did not worsen health among several lowland Indian communities of Venezuela. As villagers became part of market economies, they continued to eat game and traditional crops instead of eating canned food. Households with closer ties to the market had more cash and found it easier to buy game and traditional crops than to forage or grow crops. Households with closer ties to the market also made greater use of modern medical services.

Reasons for Divergent Views

In the debate about how markets and acculturation affect morbidity, researchers may have arrived at different positions for several reasons.

Different Forms of Integration to the Market

As discussed in chapters 3-4, indigenous people take part in markets in many ways. They sell different types of goods and services and they can rely on credit with outsiders. Different forms of exposure to the market may affect health in different ways, as discussed below.

Wage Labor

In his review of the literature, Wirsing (1985) concluded that wage labor in commercial plantations undermined the health of rural workers. Hard physical work, low real wages, and poor nutrition in a work place far from a laborer's village may combine to worsen health. However, a well-paid, blue-collar job that allows rural workers to keep a foothold in subsistence agriculture may not worsen health, although it may increase obesity and hypertension (Santos and Coimbra 1996). Wage labor may affect health in different ways, depending on the occupation.

Crops

The same ambiguity arises when rural people become part of the market by selling crops. As Kennedy (1994) showed in her international comparison, cash cropping can worsen or improve the health of rural people. The effect of cash cropping on health declines once researchers control for the type of cash crop and the amount of time spent cultivating the crop. Some cash crops absorb much time and effort, alter the local ecology, and produce large changes in how households consume and allocate resources (Behrens 1986; Burkhalter and Murphy 1989; Leonard et al. 1994a, 1994b; Murphy 1960). Other cash crops do not produce such large effects on the household or village economy (Reed 1995). The occasional sale of non-timber forest goods, for example, allows villagers to gradually modify the forest and—depending on the products sold—continue practicing subsistence farming (Anderson, May, and Balick 1991).

Credit

Depending on the type of credit market, credit will likely produce ambiguous effects on health. Incipient credit markets, with their high interest rates and bonds of debt peonage, chain indigenous people to hard toil and could worsen their health (Murphy 1956)—though the practice may confer some insurance. Research in Asia suggests that access to a well-functioning credit market improves health because it allows the sick to borrow and buy medicines, pay for health care, or hire laborers to tend their fields during bouts of illness (Amin 1997; Foster 1995). By allowing people to borrow when they get sick, credit may make it possible to keep food consumption from fluctuating and health from deteriorating (Morduch 1995; Rose 1994).

Distinguishing Between Integration to the Market and Acculturation

Most researchers have not distinguished between acculturation and integration to the market, nor have they controlled for the effect of both types of variables at the same time when studying morbidity. Wirsing (1985), for example, titled his review article, *The Health of Traditional Societies and the Effects of Acculturation*, yet in much of the review he dealt with the effects of markets rather than with the effects of acculturation. As the two earlier passages in this chapter suggest, researchers such as Wirsing, Shepard, and Rode cared about both acculturation and markets, but did not estimate the separate and simultaneous effect of each type of variable on health. Different positions in the debate may reflect researchers examining different types of variables.

Objective Health versus Self-Perceived Health

Markets and acculturation may affect health in different ways, depending on whether one defines health through self-perceived or objective criteria. Although markets and acculturation could make people healthier, they could also make people complain more about their health. Isolated or poorer villagers may need to work and may be more willing to work even when they are ill–they may therefore be more likely to under-report poor health. With greater income and exposure to the market:

- health expectations rise and people may be more likely to report illness and seek treatment (Johansson 1991; Murray and Chen 1992; Murray, Yang, and Qiao 1992)
- people may consider chronic as well as acute diseases as threats to their health
- people may worry more about more subtle aspects of health (e.g., obesity)

Hypotheses

Drawing on the debate and the literature just reviewed, one can advance the following hypotheses for empirical tests:

Hypothesis 1: Integration to the market through the sale of labor will worsen health. Greater wage income will be associated with longer absences from the village, less contact with traditional healers, and less access to traditional foods.

Hypothesis 2: Access to credit will be associated with better objective health because it will allow households to borrow during times of illness and smooth fluctuations in income and consumption. Access to credit, however, will be associated with more sick days reported because credit will allow the sick to hire people and take time off when illness strikes.

Hypothesis 3: Variables related to integration to the market will carry the same weight as variables related to acculturation in determining morbidity.

The last hypothesis does not flow from the current debate, but it is included for two reasons. First, the hypothesis forces one to estimate and compare the separate and simultaneous effects of markets and culture on morbidity—a topic latent, but not explicit, in much of the literature reviewed earlier. Second, the hypothesis touches on a larger and older debate in cultural anthropology about the relative contribution of material and non-material determinants of behavior. The third hypothesis has been phrased in a way that gives equal weight to the two positions in the debate. A hypothesis about the effects of cash cropping on health was not formulated because of spotty information on cash cropping. The first hypothesis should have made reference to how different types of occupations affect health, but such a hypothesis could not be tested because information on the choice of occupation for people working outside the village was not collected.

Description of Illness

The information collected shows that people in the sample felt they enjoyed good health. Of the 949 household heads that responded, 26 percent said they had good health at the time of the interview, 61.54 percent reported having average health, and only 11.70 percent said they suffered from poor health. Compared to how they felt a year before the interview, 24 percent of the people said they currently had better health, 61.05 percent reported enjoying about the same health, and only 14.95 percent said their health had worsened.

Subjects were asked how many days they had felt ill or had been confined to bed during the two weeks before the interview. People listed 47 different complaints or symptoms affecting their health during that period. Most of the complaints and symptoms included:

- fevers (17.93%)
- stomach pain (12.76%)
- colds (10%)
- body, head, and back pain (18.97%)
- rashes (4.48%)
- diarrhea (4.14%)
- pains in the womb (3.79%), hip, and kidneys (6.20%)

Subjects also reported having (at present or in the past) the following diseases:

- leishmaniasis (17.59%)
- dengue fever (10.58%)
- tuberculosis (5.97%)
- yellow fever (11.83%)
- malaria (17.69%)

People did not seem well protected against illness, a topic discussed at greater length in the next chapter. Virtually none of the subjects said that they had received modern medicines from outsiders—whether from loggers, cattle ranchers, colonist farmers, or traders. The exception was the Chiquitanos—64 percent reported that they had received medicines from the Catholic Church.

The sick did not seem to get support from kin or friends. Subjects were asked how many days they had been ill during the previous forest-clearing season and, if so, who had helped them during their illness. Because the area of forest that a household clears in part determines how much food it will consume over the coming year, illness during the forest-cutting season could affect consumption in a direct way. Of the 134 subjects who had been ill during the forest-cutting season, 43 percent said they had done nothing except weather the spell on their own—without help from anyone. The balance relied on labor help from their family (41.04%) or neighbors (5.97%). A few hired labor (5.22%), borrowed money from kin (3%), or sold animals (1.49%).

Definition and Measurement of Variables

The empirical analysis focuses on adults because adult morbidity and nutrition have received less attention than child morbidity and nutrition (Bailey and Ferro-Luzzi 1995:673; Strauss et al. 1993:794). Owing to limitations of space, the analysis is done for the pooled sample rather than for each separate ethnic group. Tables 9-1 and 9-2 contain definition and summary statistics of the variables used in the analysis. What follows is a discussion of how the variables were defined and measured.

Dependent Variables

Self-perceived and objective illnesses were measured separately because the second hypothesis requires that one distinguish between the two types of illness. Measures of self-perceived and objective health also did not overlap, justifying their separate treatment. Correlation coefficients between the three measures of self-perceived health and the two measures of objective health were less than 0.08. Self-perceived illness included the following:

- the number of days lost to work during the last forest-clearing season,
- the number of days ill during the two weeks before the interview, and
- the number of days confined to bed during the two weeks before the interview.

TABLE 9-1 *Definition and Measurement of Variables for People Over the Age of 16*

Variable	Definition
Dependent–	
Self-perceived:	
Ill	Number of days ill reported in past two weeks
Bed	Number of days unable to work from illness in past two weeks, as reported by subject
Forest	Number of days ill in forest-clearing season, as reported by subject
Objective:	
Blood	Dummy variable; 1=presence of blood in sputum, feces, or vomit in past two weeks; 0=no blood
BMI	Body-mass index; kg/mt^2
Explanatory–	
Continuous:	
Wage	Wage earnings/person/year in *Bolivianos* (U.S.\$1=5.23 *Bolivianos*)
Wealth	Value of physical assets/person/year in *Bolivianos*
Credit	Value of credit received/person/year in *Bolivianos*
Education	Maximum educational attainment
Math	Numeracy; 1=numerate, 0=non-numerate. Determined through a test
Age	Age of subject in years
Dep ratio	Dependency ratio (children/adults)
Nurses	Number of health professionals in villages
Distance	Kilometers from village to nearest town in straight line measured by GPS
Dummies (name=1):	
Mojeño	Membership in Mojeño ethnic group
Yuracaré	Membership in Yuracaré ethnic group
Tsimane´	Membership in Tsimane´ ethnic group
Chiquitano	Membership in Chiquitano ethnic group
Spanish	Knowledge of spoken Spanish
Read	Literacy in Spanish; determined through a test
Female	Sex of subject
Parented	Either parent had some education
Medkit	Village has medical kit

TABLE 9-2 Summary Statistics of Variables

Variable	Obs	Mean	Sd	Min	Max
Continuous:					
Ill	955	2.30	4.24	0	14
Bed	952	.982	2.63	0	14
Forest	949	1.63	6.06	0	90
BMI	955	23.8	2.71	17	41
Wage	930	182	421	0	3861
Wealth	964	557	690	2.5	6467
Credit	966	5.61	37.7	0	500
Education	956	2.69	2.74	0	16
Math	954	1.19	1.55	0	4
Age	955	38.4	13.2	14	79
Dep ratio	964	1.46	.987	0	6
Nurses	41	1.12	.640	0	3
Distance	41	93.5	37.8	48.4	186
Dummies:					
Blood	955	.204	.403		
Spanish	955	.924	.264		
Read	952	.548	.497		
Parented	955	.281	.450		
Female	966	.500	.500		
Mojeño	966	.273	.445		
Yuracaré	966	.128	.334		
Tsimane´	966	.101	.302		
Chiquitano	966	.496	.500		
Medkit	41	.731	.448		

NOTES n=41 for medical kits (medkit), nurses, and distance because these are village variables; the rest are personal variables.

The last two dependent variables differed in severity. The average subject had been confined to bed and unable to work 0.98 days (n=952; standard deviation=2.63) during the two weeks before the interview, but had felt ill 2.30 days (n=955; standard deviation=4.24). Of the three dependent variables, the first probably contains the largest measurement error since it required that people recall the length of illnesses that had occurred several months before the interview. The three definitions are subjective because those interviewed (rather than the interviewers) defined whether and how long the subject had been ill.

Objective health included the following two measures: body-mass index and the presence of blood in sputum, vomit, or feces during the two weeks before the interview. Although body-mass index is an objective measure of adult nutrition, the presence of blood may not be because we collected the information by asking subjects. All dependent variables, except for the presence of blood, are expressed in logarithms. The presence of blood was coded as a binary variable.

Information was collected on the past and present incidence of tuberculosis, malaria, leishmaniasis, dengue, and yellow fever. This information was not analyzed for two reasons (besides brevity). First, the information replicates the self-perceived measures already collected. Because the research team lacked trained medical personnel to verify the diagnoses, the diagnoses remain subjective. Second, information on these illnesses was coded as binary variables. The variables do not contain as much variance as the three earlier definitions of subjective illness, which were coded as continuous variables.

Explanatory Variables

Explanatory variables were divided into three groups:

1. variables related to integration to the market,
2. variables related to acculturation, and
3. control variables.

Integration to the Market

Integration to the market was measured in the following three ways:

1. earnings from wage labor,
2. use of credit, and
3. wealth.

For some tests, the number of nurses and medical kits in the village were also included (see control variables).

Although the results of the analysis are presented with earnings from wage labor as an explanatory variable, the results of the analysis were also done using number of days worked instead of monetary earnings. Using monetary earnings may cloud differences between different types of wage work (e.g., number of days worked inside and outside the village). The use of monetary earnings may also cloud the effort invested. Some subjects may have high wage earnings from high wages rather than long periods of work; the latter is most likely to affect health. As seen below, results hold up independent of the definition used for wage labor. The credit variable includes the value of all credit obtained inside and outside the village during the last year. Wealth is the value per person of about one dozen commercial, physical assets. All variables related to integration to the market are expressed in logarithms.

Acculturation

Acculturation was measured using the following variables:

- knowledge of Spanish
- literacy
- knowledge of arithmetic
- maximum education of the subject
- a dummy variable to capture whether either of the subject's parents had any formal education

Spanish, literacy, and knowledge of arithmetic were determined through a test (in Spanish). One relevant proxy of acculturation—with direct bearing on subjective health—is folk beliefs about health. Such information was not collected.

Each variable that proxies for acculturation or integration to the market is entered separately in the regression, but the total effect of the group of variables related to acculturation or the market is measured through a Wald test. Measuring the effect of the variables jointly allows one to avoid constructing an arbitrary index for acculturation or for integration to the market and also allows one to take the natural multicollinearity within the variables of each group into account.

Although variables related to acculturation and integration to the market overlap, they deserve separate space of their own in the same econometric model for two reasons. First, multicollinearity between the two vectors of variables was low. Pearson product-moment correlation coefficients between acculturation and market variables were estimated and all values fell below

0.40. Second, for reasons of theory discussed earlier, it was important to compare the joint effect of both types of variables on health.

Control Variables

Control variables included age, sex of subject, dependency ratio (children/adults), village attributes likely to influence health, such as the number of health workers and medical kits in the village, and distance from the village to the nearest market town. Control variables also included dummy variables for each of the three ethnic groups along the river Sécure—the excluded group was the Chiquitano.

Econometric Models, Endogeneity, and Comparing Different Metrics

This section contains a discussion of the econometric models used, potential endogeneity, and problems of metrics and comparison that arose when trying to gauge the significance of material and non-material determinants of health.

Econometric Models

A model of adult morbidity, in which the level of a person's health reflects life-cycle (e.g., age), personal (e.g., education), and household attributes (e.g., degree of exposure to different types of markets) was used (Strauss et al. 1993).

The information was analyzed using three types of econometric models. First, a tobit model was used for self-perceived morbidity because dependent variables related to subjective health were censored at zero. Seventy to 85 percent of the subjects reported no illness during the most recent forest-cutting season and the two weeks before the interview. Second, ordinary least squares with Huber White robust standard errors were used for body-mass index because the residuals for some explanatory variables displayed nonconstant variance. Last, a probit model was used to predict the presence of blood. In column 5 of table 9-4, the coefficients represent the probability of reporting blood in stools, vomit, or sputum, when an explanatory variable increases by one unit above the mean of the sample for that variable. For example, the probability of a subject reporting blood during the two weeks before the interview decreases by 1.01 percent when the subject's education rose by one year beyond the 2.54 years of education that is the mean for the sample used in the regression.

Endogeneity

Markets affect morbidity, but one could also argue that causality runs in the opposite direction. There were no instrumental variables to control for endogeneity, but one should note that most endogenous variables were lagged in time, relative to the dependent variable—thereby reducing the severity of endogeneity. Another source of endogeneity may relate to the timing of illness and credit. If households received credit after a bout of illness, illness could drive a person to seek credit rather than credit affecting a person's health. No information was collected on the timing of credit to resolve the impasse.

Comparing Different Metrics

One can do statistical tests to decide whether a group of variables related to the market (e.g., income, wealth, credit) affects health. One could do the same for a group of variables related to acculturation (e.g., literacy, fluency in Spanish). The exercise is trivial. The comparison and social interpretation of results, however, is hard (if not impossible) because the metrics used to measure the two groups of variables differ.

Suppose, for example, that a ten percent increase in income or a ten percent increase in knowledge of spoken Spanish each improved body-mass index by one percent in a well-specified, multivariate regression. Suppose further that each of the two coefficients was statistically significant and that there were no biases from omitted variables, endogeneity, and so on. One could conclude from the analysis that the two variables were equally important in a statistical sense, but not in a social or economic sense. A ten percent improvement in knowledge of spoken Spanish might require large investments, whereas a ten percent improvement in income might not, or vice versa. In another example, suppose that a five percent increase in income or a ten percent increase in a test score of knowledge of spoken Spanish each produced a one percent improvement in body-mass index (in a well-specified regression) and that the results were statistically significant. One could conclude from the estimation that the two variables did not carry equal statistical weight. The two variables, however, could carry the same economic weight if a five percent improvement in income was as hard to achieve as a ten percent improvement in linguistic competence.

These points are stressed to ensure a proper reading of the results discussed below and how they might bear on the third hypothesis. The discussion of results from the comparison of cultural and market forces focuses exclusively on the statistical and not the social or economic meaning of the coefficients.

The Limits of Bivariate Analysis: A Detour and Example

As noted previously, conventional anthropological studies of how markets or acculturation affect outcomes are likely to yield inaccurate results because they generally rely on bivariate analysis. This point is illustrated below, using health as an example.

Consider the relation between acculturation and the five measures of health discussed earlier:

- illness during the forest-cutting season
- the days the subject felt ill
- the days the subject was confined to bed during the two weeks before the interview
- body-mass index
- presence of blood during the two weeks before the interview

Assume that fluency in spoken Spanish proxies for acculturation. The column called *Bivariate* in table 9-3 summarizes the results of five separate bivariate regressions. Each regression has used a different definition of illness as a dependent variable and fluency in spoken Spanish as the explanatory variable in all regressions. The column called *Bivariate* contains information on the sign of the coefficient for the Spanish variable, the magnitude of the effect, and the t or the z value in parenthesis.

A bivariate analysis would suggest that knowledge of Spanish is associated with more self-reported illness, a four percent increase in body-mass index, and a 16.87 percent greater probability of reporting blood. Except for the results of the first regression or row, all other results were statistically significant at the 90 percent confidence level or above. A bivariate analysis would lead one to conclude that acculturation, measured by knowledge of Spanish, bore a statistically strong link to health.

TABLE 9-3 Comparison of Bivariate and Multivariate Analysis of Health (Dependent Variable) and Spanish (Explanatory Variable)

Dependent Variable	Bivariate	Multivariate
1. Illness during forest-cutting season	3.37 (1.45)	1.30 (.44)
2. Number of days ill in past 2 weeks	5.59 (3.70)	4.16 (2.20)
3. Number of days in bed in past 2 weeks	5.44 (2.72)	3.09 (1.29)
4. Body-mass index	.04 (3.24)	.03 (1.83)
5. Presence of blood	16.87% (3.37)	6.69% (.92)

NOTES Numbers reported are coefficients for knowledge of Spanish from bivariate or from multivariate regressions. Coefficients under the column called "Multivariate" come from table 9.4. t values reported in parenthesis for dependent variables 1-4; z value reported for dependent variable 5. Variable for knowledge of Spanish coded as a dummy (1=knows Spanish; 0=does not know Spanish).

Now consider the estimated coefficients for the same variable (Spanish), from the multivariate analysis of table 9-4, that are placed in the column called *Multivariate* in table 9-3. After controlling for many covariates, the effect of acculturation on health diminishes across all five econometric specifications and remains statistically significant (at the 95 percent confidence level) in only one of the five models.

Results of Multivariate Analysis

Table 9-4 contains the regression results. The table contains five columns with numbers corresponding to the five dependent variables listed on page 139.

TABLE 9-4a Determinants of Adult Morbidity

Dependent Variable	Forest [1]		Ill [2]		Bed [3]	
Variable	Coef	Se	Coef	Se	Coef	Se
Market:						
Wage	-.24	.11[2]	.04	.06	.19	.09[2]
Wealth	.90	.58	.79	.35[2]	.69	.44
Credit	-.16	.31	.24	.17	.33	.21
Nurses	-.14	1.14	-.60	.66	-.45	.84
Medkit	-2.78	1.51[1]	-1.37	.91	-.36	1.15
Acculturation:						
Education	.32	.32	-.31	.20	-.73	.28[2]
Spanish	1.30	2.91	4.16	1.89[2]	3.09	2.38
Read	1.26	1.59	.58	.93	.94	1.18
Math	-1.29	.57[2]	.39	.33	.36	.44
Parented	1.27	1.32	1.69	.78[2]	2.20	.99[2]
Observations:						
Left-censored	710		578		666	
Uncensored	116		252		161	
Total	826		830		827	
Adjusted R2	.06		.03		.03	
Joint test (Prob, %):						
Market	3.91		6.27		11.13	
Acculturation	29.33		2.87		3.50	

TABLE 9-4b *Determinants of Adult Morbidity*

Dependent Variable	BMI [4]		Blood [5]	
Variable	Coef	Se	Coef	Se
Market:				
wage	.0006	.0008	.28	.25
wealth	.003	.003	1.49	1.30
credit	.0003	.002	.49	.55
nurses	.01	.007[2]	-3.47	2.46
medkit	-.002	.01	6.58	3.11[2]
Acculturation:				
education	-.002	.002	-1.01	.73
Spanish	.03	.01[1]	6.69	5.86
read	.01	.01	-1.81	3.35
math	.003	.003	-1.99	1.24
parented	.002	.008	2.39	2.95
Observations	830		830	
Adjusted R2	.07		16.49	
Joint test (Prob, %):				
Market	33.83		14.31	
Acculturation	25.75		2.00	

NOTES Control variables (not shown) include: age, sex, dependency ratio, distance, and dummies for Mojeño, Tsimane´, and Yuracaré (Chiquitano is excluded group). Regressions include constant. Dependent variables in regressions 1-4, wage, wealth, credit, dependency ratio, and distance in natural logarithms. Regressions 1-3 are tobits, 4 is ordinary least square, and 5 is probit with robust standard errors (see text for probit). 1, 2, and 3 significant at ≤10%, ≤5%, and ≤1%. Joint test refers to the results of the Wald test (prob>F) for joint significance of variables related to market (household wage, wealth, credit, and number of nurses and medical kit in village) and for variables related to acculturation (education, Spanish, literacy, numeracy, and parental education).

Hypothesis 1

The first hypothesis states that wage labor ought to be associated with greater morbidity. The results of the analysis do not support the expectation. Wage income only exerted a statistically significant effect on morbidity in two of the five models, and the effect was not always in the expected direction. A one percent increase in wage income decreased the number of days that subjects reported being ill during the forest-cutting season by 0.24 percent (p>|t|=3.6%)(column 1), but increased the number of days of bed confinement during the two weeks before the interview by 0.19 percent (p>|t|=3.1%)(column 3). Only the latter result meshed with prior expectations.

In the other three regressions (columns 2, 4, and 5) wage income had a small physical and statistical impact on health. A doubling of wage income improved body-mass index by about only 0.06 percent (p>|t|=42.8%) (column 4), increased self-reported morbidity in the two weeks before the interview by 4 percent (p>|t|=53.1%) (column 2), and increased the probability of reporting blood by 28 percent (p>|z|=26.3%) (column 5). The models of table 9.4 were re-estimated using the following variables (instead of earnings from wage labor): 1) total days worked, 2) days worked for payment in kind and in cash, and 3) days worked inside and outside of the village. The analysis (not shown) produced the same results as those just discussed with only one exception. The number of days worked for a wage (in kind or in cash) inside the village was associated with better nutrition, but the result was significant only in a statistical sense. Doubling the number of days worked inside the village increased body-mass index by only about 0.23 percent (p>|t|=5.8%).

Hypothesis 2

Hypothesis 2 states that access to credit should be associated with better objective health and with worse subjective health. Credit should allow people to smooth consumption (and improve objective health) and take time off when ill (increasing the number of days ill reported by the subject).

Contrary to expectations or the evidence reviewed earlier, credit had no discernable effect on subjective or objective health. As expected, credit made it more likely for subjects to report illnesses and take time off when ill (columns 2-3). A one percent increase in the use of credit was associated with a 0.24 percent and with a 0.33 percent increase in the number of days reported ill during the two weeks before the interview—but in neither regression were the results statistically significant at the 90 percent confidence level or above. A doubling of credit increased body-mass index by only about 0.03 percent (column 3), but the result was statistically insignificant (p>|t|=86.7%).

Hypothesis 3

Hypothesis 3 states that variables related to the market and variables related to acculturation should carry equal weight in shaping morbidity. The hypothesis is tested in two ways—both related to statistical rather than social or economic significance.

First, the statistical weight of individual variables between two categories, acculturation and markets, are compared. If one follows this procedure, one could conclude that market and acculturation variables both affect health. In the regression of column 1, for example, two market variables (wage income and the number of medical kits in the village) decreased the number of days of reported illness during the forest-cutting season. The same regression also shows that knowledge of arithmetic (acculturation variable) had a strong, beneficial effect on health. In models 2-4, variables related to the market and acculturation both affected morbidity. In regression 3, wage income (market variable) and the education level of subjects and their parents (acculturation variables) all affected health. Only in regression 5 could one conclude that acculturation variables played no statistically significant role.

The second way of testing this hypothesis is to do a Wald test to estimate the joint statistical significance of all variables related to the market (wage income, credit, wealth, and number of nurses and medical kits in the village) and all variables related to acculturation (education, knowledge of Spanish and arithmetic, literacy, and parental education). The results of those tests are reported at the end of tables 9-4a and 9-4b, in the row called *Joint test*.

Those results again show that variables related to the market and acculturation are important as a group in a statistical sense. In the regression of column 1, market variables (but not acculturation variables) show a statistically strong joint effect on morbidity (prob>F=3.91% for market variables and 29.33% for acculturation variables). The opposite is true in column 5. In that regression, acculturation variables were, as a group, a statistically stronger determinant of morbidity than variables related to the market. The test of joint significance in the probit regression (column 5) shows that acculturation variables were significant at about the 98 percent confidence level, but that variables related to the market were statistically significant at about the 86 percent confidence level. In regressions 2-3, both types of variables mattered (as a group) at about the 90 percent confidence interval. In column 4, neither type of variable (as a group) seemed to affect health.

Conclusion

This section contains a discussion of the possible reasons for the weak statistical results for the first two hypotheses, and the contribution the chapter has made to anthropological theory and anthropological methods.

Possible Reasons for the Weak Statistical Results

The lack of strong statistical results for the first two hypotheses may have occurred for several reasons.

First, some explanatory variables may have been mis-measured and biased the estimated coefficients toward zero, or illness may have been mis-measured and raised the standard error. If both occurred, the null hypothesis of no effect would have been easier to accept. Second, most of the dependent variables (except for body-mass index) were censored at zero and had small variance. Lack of variance and measurement error in the dependent variables would weaken the statistical results. Third, the effect of markets and acculturation may be more marked and readily visible among children than among adults. By focusing on adults, it may have been harder to spot relationships. Fourth, the model may have been too crude to detect significant relations—as shown by the low adjusted R squares of table 9-4. Most of the adjusted R squares fell below 0.10. Unlike child morbidity, not enough may be known to model the multiple determinants of adult health well, especially among indigenous people in simpler economies. Fifth, using a multivariate approach may have robbed classic variables of their conventional explanatory power. A bivariate analysis of how markets and acculturation affect morbidity would have produced stronger statistical results, as shown in table 9-3. Sixth, the study was unable to control well for endogeneity. Last, it relied on cross-sectional information. A study following the same subjects over time, as they took a greater part in the market might have uncovered trends that are harder to spot in a cross-sectional study.

The Debate

As discussed earlier, policy-makers and researchers have taken at least three positions in the debate about how markets and acculturation affect the health of indigenous people. The results of the analysis presented here suggest that: 1) neither credit nor wage labor seem to exert any meaningful effect on morbidity or on nutrition, and 2) variables related to the market and to acculturation both seem to affect health in a statistical sense.

Anthropological Theory

Cultural anthropologists have long debated the importance of material and non-material determinants of behavior. Using health as a case in point, this chapter has tried to contribute to the debate by using a transparent method for estimating the relative statistical contribution of material and cultural determinants in one outcome—morbidity. The results suggest that both types of variables matter in a statistical sense and that both camps have a point.

Anthropological Methods

Hopefully, this chapter has shown that the relationship between health, acculturation, and markets should be studied in a multivariate and comparative way. Health, like many biological outcomes and forms of behavior, reflects a web of causes that includes more than markets and acculturation. Different forms of integration to the market and different dimensions of acculturation must be estimated at the same time because markets and acculturation exert a simultaneous effect on health. The estimates must be made while controlling for confounding covariates and endogeneity. Viewed this way, the chapter represents one of the first steps in the quantitative, multivariate, and comparative study of the health of indigenous people experiencing economic development and cultural change.

Chapter 10

Mishaps, Savings, and Reciprocity

Ever since the publication of Marcel Mauss' (1990 [orig. 1927]) *Essay on the Gift* early in the twentieth century, a distinguished group of anthropologists (Lévi-Strauss 1969; Parry 1993; Parry and Bloch 1989; Sahlins 1972) has stressed the prevalence of reciprocity in foraging bands and traditional rural societies and the partial replacement of reciprocal solidarity by more mechanical and impersonal forms of contracts as societies modernize (Appadurai 1986; Carrier 1990; Yang 1989).

Reciprocal relations confer many advantages. They allow people to smooth consumption, get goods and services they lack, and strengthen social bonds (Dwyer and Minnegal 1993; Goland 1993; Winterhalder 1986; Winterhalder, Lu, and Tucker 1998). Mauss (1990:2) denied a schism in the types of exchange institutions found in primitive and in modern economies, but he emphasized that gift-giving and reciprocity were more marked in earlier or simpler economies.

Since Mauss first offered his idea, anthropologists and economists have explored aspects of reciprocity, but have not yet put Mauss' idea that reciprocity weakens with modernization to a direct empirical test. Anthropologists have documented the prevalence of reciprocity in simpler economies and among the urban poor (Lommitz 1977; Smith 1996; Yellen 1990). Economists have developed models and provided indirect empirical support for Mauss. Stephen Coate and Martin Ravallion (1993), for instance, show that even self-interested people who are uncommitted to reciprocity will share, hoping that others will return favors in the future. They show that the amount of sharing based on personal motives is less than the amount of sharing from well-established norms of reciprocity and an external authority to enforce the norms (Carter 1997:564). Using a different model, Rachel Kranton (1996) arrives at the same conclusion. Marcel Fafchamps (1992) and Michael Carter (1997) develop models to show how increases in wealth and in the possibilities of drawing on one's own self-insurance weaken reciprocity. Jonathan Morduch (1995) and Mark Rosenzweig (1988) show that although poor people in rural southern India could reduce risks through more gifts and more exchanges, they show a growing preference to use credit to cope with mishaps.

Despite the many ethnographies on reciprocity since Mauss' essay first appeared, a trend and a gap have emerged in studies of reciprocity by social anthropologists. The trend consists of avoiding a quantitative approach to the study of reciprocity (Gregory 1982; Kaplan and Hill 1985:223), except by evolutionary ecologists (e.g., Bliege Bird and Bird 1997; Hames 1987; Hawkes 1992a; Hill and Kaplan 1993; Kaplan and Hill 1985; Winterhalder 1997). The trend probably started with Bronislaw Malinowski's *Argonauts* (1961 [orig. 1922]) in the 1920s and became widespread after Sahlins' (1972) rejection of formal methods to study reciprocity (Hawkes 1992b: 405). The gap consists of not having put Mauss' idea that reciprocity weakens with modernization to a direct cross-cultural empirical test. The trend and the gap are two sides of the same coin. Anthropologists have found it hard to do cross-cultural empirical tests without standardized, quantitative information from different cultures.

The evolution of reciprocity merits attention for reasons of theory and public policy. On the academic side, it is unclear the degree to which markets weaken reciprocity. Like many of the topics discussed so far, the relationship between markets and reciprocity needs to be evaluated through empirical analysis. On the public-policy side, policy-makers need to worry about the collapse of reciprocity if the collapse increases the economic vulnerability of the rural poor. Because reciprocity probably helps the poor cope better with unforeseen events, its collapse could undermine the welfare of the poor. Poor urban people enmeshed in a strong network of social support seem less likely to fall into extreme poverty (Gray-Molina et al. 1998; Moser 1998). If reciprocity is a safety net for the urban poor or for people in areas with poorly developed labor and credit markets, then the break-up of reciprocity could increase the economic vulnerability of the poor.

In a preliminary study of villages displaying different degrees of exposure to the market in northern Thailand, economist Robert Townsend found that only the villages most integrated to the market lacked

> ...internal credit and insurance arrangements of almost any kind, and for episodes of severe illness, at least, some households seem to suffer changes in consumption. This observation then raises difficult questions. *Does insurance in the form of indigenous arrangements deteriorate with [economic] growth? Should the link between insurance arrangements and growth be called into question?*
>
> (Townsend 1995:97-98; author's emphasis)

If indigenous lowland people face more volatility in consumption when reciprocal obligations disappear, as Townsend implies, governments may need to carve out a custodial role (Deaton 1997:351, 362)—although it is unclear what form that role should take. If the collapse of reciprocity increases economic vulnerability, or if people prefer to use credit or wage labor to cope with mishaps (as the evidence from south India mentioned earlier suggests), governments might improve welfare by putting policies in place that make rural financial and labor markets work better—in so doing, governments can fill the hole left by the disappearance of traditional forms of insurance (Morduch 1998).

With quantitative information from households of Tawahka, Yuracaré, Mojeño, and Tsimane´ Indians, this chapter tries to answer two questions. First, do markets weaken reciprocity, as many anthropologists and economists since the days of Mauss have said? Second, do indigenous people become more vulnerable to poverty when reciprocity breaks down? Before answering the questions one needs to discuss how to define, measure, and estimate reciprocity.

Definition, Measurement, and Estimation

Defining, measuring, and estimating reciprocity poses several challenges. First, one needs to specify the goods and services people exchange. People may share some goods, but not others; or they may share information, but not physical assets. At any one time, people may share some goods and services in an altruistic fashion, but they may hold back sharing other goods and services (Winterhalder 1996a). A study of reciprocity requires that one specify the goods and services people exchange and that need to be measured. The absence of reciprocity with some goods and services does not mean reciprocity does not exist with other goods and services. This point is documented later in the chapter.

Second, one needs to specify the time over which one will measure reciprocity (Hill and Kaplan 1993:703). A long time may elapse between the moment one receives a good or service and the moment when one returns the favor. In some societies, people and households may take many years or generations to return goods and services received as part of a longer chain of exchange. If a study covers a short time period, an item or service offered may resemble a pure gift to an outsider even though it forms part of a longer-term, reciprocal transaction.

Third, one needs to ensure that the direction of causality is specified well. Some of the classic determinants of reciprocity may be the outcome and not the cause of reciprocity (Minnegal 1996:152-153). Mauss (1990), Marshall (1961), and Lévi-Strauss (1969), among others (Aspelin 1979; Smith 1996:200), have said that reciprocity reflects social solidarity. People with strong bonds of solidarity produced by the sharing of such things as language, values, or kinship tend to exchange goods and services with each other. But one could also argue that people who exchange goods and services with regularity may, over time, develop a strong sense of solidarity, learn each other's language, adopt each other's values, and, perhaps, even create fictive kinship relationships with each other.

Last, one needs to spell out the relevant social network or parties exchanging goods and services. Reciprocity may exist within lineages, but not within moieties; or within moieties but not within all households in a village. Failure to specify the relevant social actors with care may lead one to conclude, erroneously, that reciprocity is weak or absent.

Difficulties in definition, measurement, and estimation may explain why so much of the anthropological research on reciprocity has been qualitative. Evolutionary ecologists provide an exception, and it is to their work that this text turns briefly to highlight complementarities with the approach it is about to present.

The Approach of Evolutionary Ecologists

Evolutionary ecologists have measured the flow of game, fish, or labor between (generally foraging) households, and have explained forms of reciprocity using such things as kinship links, physical proximity of partners exchanging goods, variability in encounter rates with animals (Hames 1990), seasonality of harvest (Bliege, Bird, and Bird 1997), type of food (Kaplan and Hill 1985), and—in more seasonal environments—the technology for storing food (Cashdan 1985; Winterhalder, Lu, and Tucker 1998:40).

Evolutionary ecologists have found that people who are closely related by blood, who live next to each other, and who do not have technologies for storing food are more likely to share food. Foragers are more likely to share food when they hunt individually, find large game, and find food sporadically—rather than when they find small game at the same time (Kaplan and Hill 1985; Winterhalder 1996b).

Although they have advanced our understanding of reciprocity through the application of evolutionary theory and empirical analysis, evolutionary ecologists have been less successful in controlling for third variables while estimating some parameters. The shortcoming may or may not bias the estimated parameter. The bias will depend on whether explanatory variables are endogenous or exogenous. Some of the explanatory variables of interest to evolutionary ecologists are exogenous because subjects cannot control them; the estimated parameters for those variables will be unbiased because they will remain unaffected by reciprocity or by the inclusion of third variables. Genetic distance between individuals, for example, is an exogenous variable—it may explain variance in reciprocity, but it remains unaffected by reciprocity or other variables. People cannot chose the amount of genetic distance they have with each other, although they can use kinship terms to bring genetically-unrelated people into their kin group or demote an unworthy blood kin to the status of a stranger or enemy.

Many of the explanatory variables of interest to evolutionary ecologists are endogenous because subjects have some control over them. The estimated parameter for those variables may be biased, unless researchers control for the simultaneity of decisions. Examples of endogenous variables used by evolutionary ecologists include storage and hunting technology, physical distance between exchange partners, skill, and even encounter rates with animals—a possible proxy for skill. People who reciprocate may end up with more wealth, and may be able to acquire better storage and hunting technology. Perhaps good hunters need to get better storage technology because they engage in reciprocity. People who swap goods and services with regularity may want to live close to each other to lower transaction costs. People who engage in reciprocity may end up having better nutrition (Kaplan and Hill 1985) and thus become better hunters and live longer. Agricultural productivity may determine the amount of goods partners exchange (Cashdan 1985:464-465), but the density of a person's exchange network may dictate how much a person needs to produce and store if she or he wishes to maintain the network. Evolutionary ecologists have not always explained how many of their explanatory variables got there. Physical distance between households, storage technology, agricultural productivity, and skill are all variables over which subjects have some choice and may reflect reciprocity. They cannot be taken as given.

Although some evolutionary ecologists (e.g., Hill and Kaplan 1993:701) acknowledge problems of endogeneity and omitted-variable bias, many have not corrected for the biases for several reasons. First, much of the work of evolutionary ecologists has emphasized the biological or ecological features of the resource (e.g., seasonality of prey) in shaping reciprocity. Because these variables are exogenous or random, the parameters estimated from a bivariate analysis can stand on their own. Second, to get more accurate estimates of endogenous variables requires the use of slightly more advanced statistical techniques (e.g., instrumental variables). Many of these techniques have not become part of the tool kit of evolutionary ecologists.

The advantage of the approach of evolutionary ecologists over the approach presented in this chapter is that evolutionary ecologists provide more accurate measurement of reciprocity between households and people. Through direct measurement of the goods and services flowing between parties, evolutionary ecologists provide high resolution of the dependent variable. In contrast, this study uses multivariate analysis, controls for endogeneity and for the role of third variables, and draws on a large sample, but measures reciprocity in an indirect way.

A New Approach to Reciprocity

This section presents a new way of looking at reciprocity—one that overcomes some of the hurdles just discussed. Reciprocity is analyzed by measuring changes in the wealth of a household between two periods of time (t and t+1), and relating the changes to the random misfortunes of one's neighbors at time t. If people practice reciprocity in times of need, then the random mishaps of one's neighbors should reduce one's stock of wealth. The study, in particular, tests whether illness in all other households of the village causes my wealth in domesticated animals to decline. The focus is on savings in animals rather than on savings in other physical assets (e.g., manioc) for reasons discussed later.

Although the misfortunes of a neighbor may not affect one's savings in animals, one may still engage in reciprocity with his or her neighbor—by exchanging goods which neither can get alone, by exchanging an occasional day in the field, or by giving each other a plate of cooked food or a piece of raw game meat when surplus is available—but such tit-for-tat reciprocity will not serve as a perfect surrogate for insurance.

The Tawahka illustrate how reciprocity might occur every day, but might not gain prominence during misfortunes. As discussed later, information from the 1995 household survey did not support the hypothesis that the mishaps of a neighbor triggered reciprocity. Households did not reduce their savings in domesticated animals in response to their neighbor's illnesses. However, an analysis of all the types of goods (n=232) entering households during weigh days over two years (1995-1996) showed that 9.98 percent of all goods entered as gifts. Goods received as gifts included crops (29%), game meat and fish (16.40%), and fruits (7.61%), followed by many types of commercial goods and forest plants. The example shows how a study of reciprocity in the same culture during the same time can yield two different results—depending on whether one measures reciprocity as the immediate transfer of goods to a neighbor following an episode of bad luck (in which case there was little or no reciprocity) or as the ordinary flow of goods between households over time (in which case there was reciprocity).

When reciprocity serves as insurance, the misfortunes of others should trigger redistribution of assets toward them. One's assets, however, could be tied to the misfortunes of a neighbor in other ways besides reciprocity. Suppose an epidemic causes most households in a village to lose their animals at the same time, making it hard to borrow from each other. Then, even if people did not practice reciprocity, one would see a link between the decline in the assets of a household and the misfortunes of neighbors. Suppose a household loses animals from mismanagement, but rationalizes the loss as having arisen from mishaps. Even if there was no relationship between the mishaps of a neighbor and a change in my stock of animals, one might conclude that the mishaps of the neighbor caused my savings in animals to decline. Another link may come through correlated errors in measurement such that a neighbor's shock may be an indicator of one's own poorly measured shock. There may be few ways to handle the last point, but the first point can be addressed by controlling for one's own shocks besides controlling for the shocks of one's neighbors.

A Reduced-Form, Unrestricted Model of Savings

Studies in Western societies and rural Thailand have shown that people tend to save out of permanent income more than out of transitory income—although both matter (Gersovitz 1988; Paxson 1992). For people in rural economies, permanent income corresponds to attributes such as landholding, physical assets, and human capital. Transitory income corresponds to attributes affecting consumption or savings in more immediate ways, such as unexpectedly good or bad rainfall and highs and lows of agricultural production over time (Deaton 1997:351-352).

The empirical analysis expresses household i's savings (Si) as the change in the value of a portfolio of domesticated animals between two years—t and t+1:

$$S_i = \delta_{0i} + \delta_1 X^o_{ti} + \delta_2 X^n_{ti} + \delta_3(X^o_{ti} * X^n_{ti}) + \delta_4 Z_{ti}^n + \delta_5 V_{ti} + \mu_i \qquad \textbf{(EQ 10-1)}$$

Si represents the change in the value of the animal stock of household i between time t and t+1. The constant δ_{0i} captures fixed effects of the household. The variable X^o_{ti} measures the shocks household i suffered during time t and proxies for transitory income. X^n_{ti} captures the average shocks to neighbors in the rest of the village (excluding household i) during time t. Z^n_{it} is a vector of household characteristics associated with permanent income at time t. The term $X^o_{ti} * X^n_{ti}$ captures the interaction between the misfortunes of household i and the misfortunes of all other households in the village at time t. V_{ti} is the average savings in animals in the village, excluding the savings of household i at time t (Townsend 1995). μ_i is an error term, or unexplained household savings.

With reciprocity, the coefficient δ_2 of my neighbor's misfortune should be negative. If Mauss and his heirs are right, one ought to see a greater reduction in one's own savings in response to a neighbor's shock among households in relative autarky, than among households in villages with closer ties to the market. In more modern villages, people ought to be able to draw on credit and other forms of insurance during bad times besides drawing on the good will of their neighbors. Thus, the coefficient δ_2 should be smaller (less negative) among dwellers of modern villages than among dwellers of villages farther from the market.

Testing reciprocity through this reduced-form savings model has two advantages. First, it allows one to mitigate potential biases from endogeneity because subjects cannot influence the random misfortunes falling on their neighbors. It also allows one to test the degree to which indigenous lowland populations save out of permanent or transitory income. If savings come chiefly from transitory income, one could conclude that people even in these relatively isolated societies are setting aside some of the dividends from good years to even out fluctuations in consumption during bad years (Deaton 1997:351-353).

Ethnographic Context of Misfortunes and Coping Mechanisms

The misfortunes affecting the ethnic groups and the coping mechanisms they used are described next.

Types of Misfortunes

Households in the sample faced many types of blows. Misfortunes included losses of crops from theft, pests, diseases, and bad weather. The 101 Tawahka households surveyed in 1995 said that in 1994—a typical agricultural year—they had lost about a third of their potential bean and rice harvest and 55 percent of their cacao harvest to unexpected events. During the floods of 1992-1993, the Tsimane´, Mojeño, and Yuracaré households interviewed said they lost 30, 60, and 47 percent of their rice harvests. During 1997—a normal agricultural year—Mojeño and Yuracaré households lost 39 and 23 percent of their rice harvests.

Besides crop losses, households face other types of misfortunes, including fires, theft, predation from wild animals, and illness. Nine percent of the Mojeño and the Yuracaré households interviewed remembered losing a

TABLE 10-1 Loss of Domesticated Animals from Predation and Theft: Mojeño and Yuracaré, 1997

Animal	Mojeño		Yuracaré	
	Obs	%	Obs	%
Cattle	36	13.88	17	1.91
Ducks	68	37.09	37	38.67
Chickens	116	31.09	57	21.67
Pigs	61	23.06	33	25.47
Dogs	103	6.82	58	3.24

NOTES Percent is share of animal stock owned one year ago lost to predation of wild animals and to theft in past 12 months.

home to fire. As the information in table 10-1 shows, theft and predation by wild animals during 1997 caused Mojeño and Yuracaré households to lose 21-38 percent of their ducks, chickens, and pigs. Losses of hunting dogs (3.24-6.82%) and cattle (1.91-13.88%) were smaller.

Households also confronted illness—another type of shock. As discussed in the previous chapter, illness during the forest-clearing season affects how much rainforest households can cut, and how much they can plant and consume during the coming year. This study tested for the correlation between the average amount of illness in a household and the average amount of illness in all other households in the village, and found low Pearson product-moment correlation coefficients (Tsimane´ r=-0.07; Tawahka r=0.03; Mojeño r=0.16; Yuracaré r=-0.06). The low values support the observation of Townsend (1995:84-85) that in rural areas of poor countries the fate of different households in a village may not move in unison.

Coping Mechanisms

To cope with shocks, the Tsimane´, Mojeño, Yuracaré, and the Tawahka rely on outside institutions and the indigenous government of their ethnic group. Government and non-government organizations have been moving in to meet the needs of indigenous people when regional calamities take place. During the floods of 1992-1993—which struck all four cultures under study—and during the flood of 1997 (which struck only in Bolivia), the groups received food aid from the central government and from non-governmental organizations. The government of each indigenous group has started to help families cope with some types of misfortunes. The Federación Indígena Tawahka de Honduras, for example, sometimes pays for the transport and treatment of the Tawahka in need of emergency evacuations—in coordination with the local church. Until recently, the Gran Consejo Tsimane´ used royalties paid by logging firms to cover the medical expenses and transport costs of Tsimane´ treated at the hospital in the town of San Borja. Coverage has declined in recent years from the decline of logging.

Although they have access to insurance from their indigenous government and outside institutions, households do not rely on formal institutions to survive. Households also take steps of their own to either ward off calamities or cope with them once they occur. Steps taken after mishaps strike include going to work for a wage in logging camps, obtaining credit from outsiders, tightening their belts, stealing, receiving help in goods or services from other villagers, or using their own savings in physical assets (Godoy, Jacobson, and Wilkie 1998).

Table 10-2 tabulates the responses of Mojeño and Yuracaré households to evaluate how they coped with different types of mishaps.

TABLE 10-2 *Mechanisms for Coping with Mishaps: Mojeño and Yuracaré (Percent of Households Using Different Strategies)*

| Mishap | Family | | | Neighbors | | Wage Labor | Forest | | | Noth | Oth |
	Obs	Borr	Labor	Borr	Labor		Plant	Game /Clear	Animal Sale		
Deaths	37	10.81		2.70	2.70	8.11			13.51	62	
Emergency	30	6.67		6.67		6.67	3.33		20		56
Rice loss	108					6.48	.93		3.70	22	42
Fires	16		12	6.25	12.5				12.50	31	25

NOTES   Mishaps are deaths in household during 1997, emergency medical evacuations in 1997, rice losses in 1997, and fire in the house. Borr (borrow) and labor are resources borrowed or labor received, either from family or neighbors. Plant, game, and clear refers to whether or not households collected/extracted plants (e.g., timber), hunted, or cleared more forest in response to a shock. Noth means household did nothing, and Oth is other strategy. Rows may not add up to 100% because of rounding. Rice loss excludes 22% who said they coped by switching to other foods.

Rows contain a summary of information about four types of mishaps: 1) deaths in the family during 1997, 2) medical emergencies during 1997,

3) rice losses during 1997, and 4) fires. Columns contain different types of coping mechanisms, such as working for a wage, receiving help from kin and neighbors, or doing nothing. By reading down a column, one can judge the importance of an institution or social arrangement in helping households cope with different types of shocks. By reading across a row, one can see the different arrangements households use to adjust to any one type of mishap. The information in table 10-2 suggests that 22-62 percent of households weathered shocks on their own—without help from family or neighbors. Households relied on wage labor and the sale of domesticated animals to cope with misfortunes.

Table 10-3 presents information about the share of different types of domesticated animals given away as gifts or lent to villagers over the last 12 months. The information suggests that households gave chickens and hunting dogs as gifts, more than cattle, ducks, or pigs. The average household gave about 9 percent of its chickens and 15-19 percent of its hunting dogs as gifts, but they gave less than 3 percent of their more valuable animals. Households almost never lent domesticated animals.

TABLE 10-3 *Share of Animal Stock Given as Gift or Lent in Last 12 Months: Mojeño and Yuracaré, 1997*

	Mojeño			Yuracaré		
Animal	Obs	%Gift	%Lent	Obs	%Gift	%Lent
Cattle	36	.29	.1	17	0	.3
Ducks	68	1.06	0	37	3.01	0
Chickens	118	9.46	0	57	8.99	0
Pigs	61	1.63	0	33	.43	0
Dogs	103	19.01	0	58	15.66	0

NOTES   Obs is number of households answering the question. Gift and lent are share of animals given as gift or lent to neighbors in last 12 months.

Savings in Domesticated Animals and Misfortunes

Like other lowland Indians, the Tawahka, Mojeño, Yuracaré, and Tsimane´ save in manioc (Carneiro 1983), trees (Irvine 1989; Posey 1984), and domesticated animals. Savings in animals rather than in other physical assets are used for two reasons. First, animals are cheap to acquire, grow in value, tolerate neglect, are easy to transport, and enjoy a well-established demand in the market. Second, people do not remember well the amount of manioc they have underground, or the total number of different types of trees they have, so it is hard to estimate the value of savings in other assets with accuracy. The use of animals to measure savings could produce error in estimation if needy people draw on other forms of savings (e.g., seeds) besides domesticated animals (Rosenzweig and Wolpin 1993; Udry 1995).

The Tawahka, Mojeño, Yuracaré, and Tsimane´ save in different types of domesticated animals. Chickens and dogs were kept by over 64 percent of the households, followed by pigs and ducks at about 45 percent (see table 10-4). Many fewer households had cattle (13-29%). The information in table 10-5 suggests that households in the four cultures kept a diversified portfolio of animals. In Bolivia, less than 20 percent of households had one or fewer animal. Three quarters of the households in Bolivia and 43 percent of the households in Honduras had 2-4 different types of domesticated animals.

TABLE 10-4 *Share of Households Owning Different Types of Animals*

Animal	Mojeño (Obs=132)	Yuracaré (Obs=62)	Tsimane´ (Obs=209)	Tawahka (Obs=101)
Cattle	28	22	13	29
Pigs	47	45	35	21
Ducks	40	45	19	n/a
Dogs	70	64	76	n.a.
Chickens	88	95	87	90

NOTES Obs is number of households answering the question.
Data indictes percentage of households.
Gift and lent are share of animals given as gift or lent to neighbors in last 12 months.
n/a = not applicable
n.a. = not available

TABLE 10-5 *Composition of Animal Stocks Owned by Household*
(Percent of H'ouseholds)

Different types of animals owned by household	Mojeño (Obs=132)	Yuracaré (Obs=62)	Tsimane´ (Obs=209)	Tawahka (Obs=101)
None	2.27	0	2.39	7.92
One	11.36	19.35	14.35	48.51
Two	25	22.58	46.89	37.62
Three	33.33	30.65	24.40	5.94
Four	25.76	20.97	9.09	0
Five	2.27	6.45	2.87	0
Total	100	100	100	100

People keep a portfolio of different animals for several reasons. First, people use small animals (e.g., chickens) to meet small emergencies. The smaller domesticated animals have become the preferred form of savings for the poor (and for single women) because they are easy to buy and sell. As household wealth increases, people save in larger domesticated animals (e.g., cattle) (Godoy et al. 1996). Second, keeping a portfolio of different types of animals probably increases the security of total savings. Small animals have high rates of reproduction but also have high vulnerability, whereas larger animals increase in size more slowly but have lower risks (Mace 1990; Mace and Houston 1989).

Definition and Measurement of Variables

Tables 10-6 and 10-7 contain definition and summary statistics of the variables used in the analysis.

Dependent Variable

For the Tawahka the study used the change in the value of a household's stock of chickens, pigs, and cattle between 1994 and 1995 to measure savings. For the Tsimane´, Mojeño, and the Yuracaré the study measured the change in the value of cattle, pigs, chickens, and ducks between 1995 and 1996 (Tsimane´) or between 1997 and 1998 (Mojeño and Yuracaré).

TABLE 10-6 Definition of Variables

Variable	Definition
Savings	Change in the value of animals between 1994 and 1995. For Tawahka: pigs, cattle, chickens. For Bolivia: pigs, cattle, chickens, and ducks
Own illness (oillness)	Number of days ill during bean harvest season (Tawahka) or forest-cutting season (Tsimane´, Mojeño, and Yuracaré)
Village illness (villness)	Average illness in households of rest of village, excluding subject's household
Interact	Interaction of own and village illness (oilness and villness)
Village savings (vsavings)	Average saving in animals of households in village excluding subject's household
Age	Age of subject in years
Household size (hhsize)	Total number of household members
Education	Maximum formal educational attainment of subject
Land	Total land holdings of household in hectares

Explanatory Variables

Information on the household's total land holdings under fallow and under cultivation, formal educational attainment, household size, and the age of each person in the household was collected.

TABLE 10-7 Summary Statistics of Variables

Variable name	Tawahka			Mojeño			Tsimane´			Yuracaré		
	Obs	Mean	Sd	Obs	Mean	Sd	Obs	Mean	Sd	Obs	Mean	Sd
Savings	783	2116	4176	264	-528	1419	740	330	1590	124	-674	1464
Oillness	769	1.73	7.70	260	1.20	3.81	463	6.5	15	117	.51	2.17
Villness	783	13.3	6.83	264	2.35	2.31	736	13	5.9	124	1.01	.93
Vsavings	783	1817	1119	264	-554	578	735	337	999	124	-596	443
Age	781	16.4	15.2	260	39	13.5	437	31	12.9	117	36.6	12.9
HHsize	783	9.18	3.27	264	6.4	2.2	736	5.05	3.08	124	6.3	2.46
Education	783	2	2.36	261	2.3	2.08	437	1.44	2.26	117	1.78	1.86
Land	780	2.52	9.85	262	3.1	1.93	712	18.3	40.4	124	3.0	1.92

Shocks were measured by the number of days the subject was too ill to work during the bean season of 1995 (Tawahka), and during the forest-cutting season of 1995 (Tsimane´) and 1997 (Mojeño and Yuracaré). A variable was constructed for the average amount of household illness in the village, excluding the household of the subject. The average amount of animal savings in the village was also estimated, excluding the savings of the subject's household. A term was included that captured the interaction of the illness of the subject and the rest of the village. All variables are expressed in logarithms.

To test Mauss' idea that markets erode reciprocity, the sample was split, using distance to the nearest market town as a proxy for participation in the market. For the Tawahka, the sample was split between the village of Krausirpe (modern) and the other villages farther up the river. For the Tsimane´, 25 kilometers from the town of San Borja was the distance used to divide the population. For the Mojeño and for the Yuracaré, 100 kilometers from the city of Trinidad was used.

Results

Results are presented in two parts. The first part contains a discussion of whether markets weaken reciprocity. The second part contains estimates for deciding whether households save out of transitory income, and whether the tendency to put aside income in good years to weather bad times grows stronger as households move closer to the market. Table 10-8 contains the regression results.

Markets and Reciprocity

The results of tables 10-8 and 10-9 show that most of the signs of the coefficients for community shocks (villness) are positive and not negative—as one might have predicted from a reading of Mauss. Except for the Yuracaré closest to the market, the misfortunes of a neighbor are associated with an increase in the savings in domesticated animals of my household. Among households closest to the market, for example, a doubling in the level of illness in the rest of the village produced increases in a subject's savings in animals that ranged from about 331 percent (Tawahka) to 24-63 percent (Mojeño and Tsimane´). Although community shocks produce large effects on savings, they did not produce statistically significant results at the 90 percent confidence level or above.

TABLE 10-8 *Effect of Markets on Reciprocity: Households Close to Market*

Variable	Tawahka		Tsimane'		Mojeño		Yuracaré	
	Coef	Se	Coef	Se	Coef	Se	Coef	Se
Oillness	1.40	3.06	-.56	.72	.14	.22	-.04	.12
Villness	3.31	5.30	.63	1.29	.24	.34	-.08	.24
Interact	-.58	1.16	.24	.29	.04	.08	-.02	.05
Vsavings	-85	9.45[3]	-1.77	1.09[1]	-.78	.52	-2.25	.79[3]
Age	.02	.12	.16	.84	.85	1.47	-2.07	.70[3]
HHsize	.53	.35	-.76	.47	.11	.91	-.16	.63
Education	.09	.10	.20	.09	.12	.17	-.23	.10
Land	.05	.07	-.28	.19	-1.33	.66	.29	.15
Obs	465		110		52		62	
Adj R2	.18		.10		-.02		.21	
F test (prob>F)	30.08%		3.31%		22.94%		.91%	

Only among the Yuracaré closest to the market does one find the hypothesized negative relation between a neighbor's shocks and one's reduction of savings, but the economic and statistical magnitude of the response was low. A doubling in the amount of illness in the rest of the village reduced the stock of savings in a subject's household by about only 8 percent and the result was statistically insignificant (p>|t|=73.9%).

If one compares the size of the coefficient for community shocks in villages far from the market with villages close to the market, one finds that the coefficients do not differ by much. The elasticities of savings with respect to community shocks are 0.21 (p>|t|=52%) and 0.24 (p>|t|=47%) for the Mojeño living far and close to the market. The comparable elasticities for the Tsimane' are 0.63 (p>|t|=62%) and 0.66 (p>|t|=19). Among the Yuracaré, the coefficients have the wrong signs. People in remote villages increase their savings when their neighbors suffer a shock (elasticity of 0.17; p>|t|=58%), whereas those living closer to the market reduce their savings (elasticity of −0.08; p>|t|=73.9%).

Only among the Tawahka does one find that villagers in more remote settlements responded as Mauss might have predicted, but even then the response was positive—not negative. The increase in savings associated with a neighbor's shock among households in remote settlements was lower (elasticity of

TABLE 10-9 *Effect of Markets on Reciprocity: Households Far From the Market*

Variable	Tawahka		Tsimane'		Mojeño		Yuracaré	
	Coef	Se	Coef	Se	Coef	Se	Coef	Se
Oillness	.24	.26	.01	.36	-.01	.09	.14	.25
Villness	.59	.51	.66	.51	.21	.32	.17	.31
Interact	-.20	.10	-.01	.14	.08	.07	.07	.07
Vsavings	-4.20	.99[3]	.83	.21[3]	-.09	.48	-.12	.60
Age	.41	.22[1]	-.40	.60	-1.55	.59[3]	.90	.87
HHsize	2.00	.65[3]	.65	.36[1]	-.58	.51	.06	.67
Education	-.13	.19	-.006	.08	-.04	.08	.01	.10
Land	-.03	.14	-.40	.18[2]	-.31	.20	-.67	.57
Obs	299		284		206		55	
Adj R2	.10		.07		.06		-.08	
F test (prob>F)	1.14%		11.47%		.31%		71.35%	

NOTES Dependent variable is savings in animals.
Regressions are ordinary least square with constant; regression weighted by household size. F test is for joint statistical significance of variables related to permanent income: age, household size, education, and land holding. For definition of far and close, see text. All variables expressed in logarithms. 1, 2, and 3 significant at ≤10%, ≤5%, and ≤1%.

0.59; p>|t|=24%) than the increase in savings among households in settlements closer to the market (elasticity of 3.3; p>|t|=53%). The response was not statistically significant in either traditional or modern Tawahka settlements.

The analysis does not appear to support the common idea that economic modernization weakens reciprocity. Most of the relations between village misfortunes and one's savings were positive—not negative—and statistically insignificant. Villagers far from and close to the market showed similar elasticities of savings with respect to the misfortunes of their neighbors.

Saving Out of Transitory or Permanent Income

To test whether people saved out of permanent or transitory income, the joint statistical significance of the variables associated with permanent income—landholding, education, age, and household size—were examined. The results were compared with own illness—the instrumental variable for transitory income.

The results suggest that households save out of permanent income, whether they live close or far from the market. For people in remote communities, results of the F test were statistically significant at about the 90 percent confidence interval for all groups except for the Yuracaré. For people in communities closer to the market, results were statistically significant among the Tsimane´ and Yuracaré at about the 95 percent confidence interval. The results do not appear to support the idea that indigenous people in the lowlands put resources away for a rainy day. In none of the regressions did the instrumental variable for transitory income—own illness—influence savings in a strong way.

Conclusion

Before turning to the questions raised at the start of the chapter, the shortcomings in the information and methods used will be reviewed and summarized.

The results are tentative because illness may have been the wrong shock to examine reciprocity. Perhaps households give animals to their neighbors, but only when other misfortunes (e.g., crop loss) occur. Second, the wrong type of dependent variable may have been chosen to examine reciprocity. Perhaps people help the needy and sick, but they do so by helping in the field, by giving food, or by lending a friendly ear, not by giving away animals. Perhaps households do give animals away to the sick, but only if the sick belong to a narrowly defined kin or residential group smaller than the entire village.

Bearing these caveats in mind, it is time to return to the original queries of the chapter—to some of the methodological concerns raised earlier—and assess the advances made.

Markets and the Demise of Reciprocity

Weak support has been found for the idea that markets erode reciprocity. The positive sign between the misfortunes of one's neighbors and one's own savings suggests that the transition out of a gift and exchange economy—if ever there was one—has already taken place. Perhaps because they lack other forms of insurance, the sick may be willing to pay their neighbors with animals and increase their neighbor's savings in physical assets.

Though cast in a different mold, the studies of Townsend (1995:96-97) in rural northern Thailand and south India, the study of Udry (1995) in northern Nigeria, and the recent review of Morduch (1998) echo some of the findings of this chapter. Townsend found that households did not reduce real capital assets to smooth fluctuations in income. They relied on crop inventories, credit, wage labor, and cash, instead. Richer households were more likely to use credit, whereas poorer households were more likely to sell labor or to use buffer stocks. In northern Nigeria, Udry found that households were generally poorly insured, and were more likely to use grain inventories than animals to cope with income shocks. He also found that the availability of credit made it easier for households to protect themselves against income shocks (Udry 1994). In a recent review, Morduch (1998) finds evidence to suggest that households in developing countries remain poorly insured against most idiosyncratic shocks.

The Demise of Reciprocity and the Welfare of Indigenous People

Because evidence that markets weakened reciprocity was not found, it is difficult to answer the question posed at the outset of whether markets increase economic vulnerability. The evidence suggests that households did not seem to set aside resources during good years for use during bad years. One may, therefore, infer that consumption might fluctuate in response to income shocks.

The absence of evidence for the use of reciprocity to cope with mishaps raises questions about the custodial role of government. Evidence has been presented elsewhere that suggests that the intensification in the use of old-growth rainforest seems to be an important way of coping with unforeseen shocks among the Tsimane´ (Godoy, Jacobson, and Wilkie 1998). The finding is supported by many ethnographies from Africa (Falconer and Koppell 1990:88; Falconer and Arnold 1989; Townson 1994) and Latin America

(Ogle 1996; Scoones, Melnyk, and Pretty 1992). These studies have shown that forests act as a shock absorber when idiosyncratic and more generalized calamities strike rural households. Old-growth rainforests are like a savings account—ready for use on short notice. If more quantitative studies confirm the idea that the use of old-growth rainforest is the principal safety net of lowland indigenous people, then the custodial role of the government falls neatly into place. Governments should protect the property rights of indigenous people to natural resources if they wish to reduce the economic vulnerability of indigenous people.

Lessons in Methods and Theory

This chapter has tried to contribute to both methods and theory in the anthropological study of reciprocity. On the methodological side, it has contributed in at least three ways. First, the study controlled for endogeneity by examining how random shocks to neighbors might affect reciprocity. Second, a multivariate approach that controlled for personal, household, and village attributes (and some of their interactions) at the same time was used. Third, a relatively large sample size was used to test hypotheses.

On the theoretical side, the idea that reciprocity is motivated by the need to help those in distress was analyzed and weak support was found for the belief. A household model of savings was used to test the prevalence and the weakening of reciprocity. The results of the analysis suggest that breakthroughs in anthropological studies of reciprocity will come from sharper theories, larger samples of households and cultures, and from more careful empirical work.

Chapter 11

Trade and Cognition:
On the Growth and Loss of Knowledge[1]

People have often lamented the loss of plant and animal knowledge by indigenous people as they join the market (Diamond 1994; Goleman 1991; Plotkin 1993; *Time* September 23, 1991). Knowledge of cultivars and forms of tillage that grew at a geologic tempo over centuries may be vanishing, as rural people adopt modern plant varieties and new ways of farming.

Although markets may erode knowledge of some plants and animals, they are also likely to produce greater retention and more knowledge of other plants and animals. Recent studies show that markets exert more complex and unclear effects on such things as genetic erosion of crops (Bellon and Taylor 1993; Brush, Taylor, and Bellon 1992) and biological diversity (Wilkie and Godoy 1996), than previously thought.

Markets may also affect the knowledge of the forest plants and game of indigenous people in different ways. Knowledge varies within and across villages (Boster 1984; Browner 1991; Brush 1993:663; Garro 1986; Kaja 1984; Scoones and Thompson 1994; Sillitoe 1998:232-233), and it is unclear how taking part in the market economy helps to explain that variation. This chapter draws on a Ricardian trade model to argue that markets are associated in system-

---

[1]Chapter written with Nicholas Brokaw, David Wilkie, Daniel Colón, Adam Palermo, Suzanne Lye, and Stanley Wei.

atic ways with the loss of knowledge of some forest plants and game and with the retention of knowledge of other forest plants and game. To test the idea, it draws on fieldwork among the Tawahka in the villages of Yapuwás and Krausirpe.

A Ricardian Trade Model and the Loss of Knowledge

As discussed in chapter 2, "Comparing Approaches," communities opening up to trade with the rest of the world ought to specialize in producing those goods in which they enjoy a comparative advantage (Krugman and Obstfeld 1997). People in isolated villages of the rainforest pursue many economic activities at the same time. They hunt, fish, farm, and collect wild plants. As trade with the outside world increases, villagers concentrate their efforts on the production of only a few goods. People begin to specialize in hunting, collecting, or fishing (Winterhalder and Lu 1997:1361). Within each of these activities, people specialize in producing those goods that they are better at producing. People use the increased output from specialization to exchange with outsiders for 1) goods that did not enter the village economy, or 2) goods that absorbed too much effort before trade took place.

The Ricardian principle of comparative advantage—the idea that people or villages specialize in those activities at which they are best—explains specialization, interdependence with the outside world, and mutual gains from trade. More importantly, the Ricardian trade model helps to generate hypotheses about what might happen to the knowledge of flora and fauna as villagers start to trade with the outside world. If people specialize in a few activities as their communities penetrate the market, they will learn more and more about those activities. Adam Smith underscored the effects of trade on knowledge. He said:

> Men are much more likely to discover easier and readier methods of attaining any object, when the whole attention of their minds is directed towards that single object, than when it is dissipated among a great variety of things. But, in consequence of the division of labor, the whole of every man's attention comes naturally to be directed towards some one very simple object. It is naturally to be expected, therefore, that some one or other of those who are employed in each particular branch of labor should soon find out easier and readier methods of performing their own particular work....
>
> (Smith 1884:5)

Smith meant that a person's knowledge of an activity increases in direct correlation to the amount of time spent on that activity. Smith implied that with increasing specialization also came a narrowness of outlook that was less marked before societies opened up to trade. The process of learning-by-doing has been documented in many societies (Audretsch and Feldman 1996; Boya 1996; Camerer, Loewenstein, and Weber 1989), including rural societies of developing countries (Foster and Rosenzweig 1995). Despite years of research on the topic, scholars have still not been able to separate the one-time effects of the beginning of trade from the more dynamic effects of trade over time. This chapter does not have information to test dynamic effects, but instead points to some correlations between trade and knowledge.

It follows from a Ricardian trade model that, depending on the type of economic integration to the outside world, markets may either enhance or erode knowledge of plants and game. Integration through the market for annual crops or labor signals a shift away from foraging and, consequently, implies that people will probably lose knowledge of wild plants and game because they will spend less time looking for them.

If integration to the market, however, takes place by selling goods from the forest, people may end up knowing more about forest goods exported from the village. When compared with those in more isolated settings, indigenous people with a stronger toehold in a market for forest goods may know more about fewer kinds of plants and game. Put differently, indigenous people in isolated settings with weak contact to the market should have generalized knowledge of flora and fauna, whereas indigenous people with tighter links to the market should have more specialized knowledge of those specific forest goods entering commercial channels.

Econometric Model

The knowledge an indigenous person retains about the plants and animals of the rainforest reflects attributes of the person (e.g., age), the household (e.g., degree and type of integration to the market), and the village (e.g., distance from village to town). Knowledge of plants or animals, Y, of subject i in household j and village k is expressed as:

$$Y_{ijk} = \alpha + \beta X_{ijk} + \Phi Z_{jk} + ?V_k + e_{ijk} \qquad \text{(EQ 11-1)}$$

where,

Y_{ijk} = knowledge of flora or fauna of person i in household j and village k

a   = intercept or personal fixed effects

X_{ijk} = vector of attributes of subject i in household j and village k

Z_{jk} = vector of attributes of household j in village k

V_k = village dummy to capture fixed effects of village k

e_{ijk} = unexplained knowledge or random error term

β, Φ, and ? are the coefficients one needs to estimate.

The econometric model builds on the spirit of James Boster's (1986) work, but differs from it in three ways. First, it controls for village fixed effects. A village with more biological diversity may have people who know more about flora and fauna because they have greater exposure to a broader range of plants and animals. Second, it controls for socioeconomic covariates (e.g., wealth) as well as kin-group composition, age, and sex. As noted, the Ricardian trade model suggests that a household's degree and type of integration to the market, or the amount of economic specialization it pursues, should affect test scores. Last, it tries to control for endogeneity by using lagged explanatory variables or variables that occurred before the dependent variable.

Methods

Two methods were used to collect information. First, during 1994-1995 socioeconomic and demographic information was collected on explanatory variables. Later, in 1996, a test was applied to elicit knowledge of forest plants and game. The information from the test was used to construct the dependent variables.

Test of Knowledge

During July of 1996, 36 Tawahkas in the village of Krausirpe and 44 Tawahkas in the village of Yapuwás—all over the age of five—were asked eight questions about forest plants and seven questions about forest animals (see appendix). Subjects came from a subset of families in Krausirpe, and included all the people present in Yapuwás at the time of the test. Subjects were part of the sample that had been monitored since the middle of 1994.

The questions that were chosen varied in difficulty to get variance in test scores. The plants and animals included in the test were a mix of used and relatively unused species. About six of the 14 species in the plant test were used infrequently (two species appeared twice), and four of the 13 animal species in the test were used frequently.

One of the two tests was selected at random for each subject. The appendix contains the questions in each test. Subjects were asked to identify specimens from illustrations in books, and to answer questions about the ecology of plants and the behavior of animals. Subjects had 30 seconds to answer each question and could answer in Tawahka, Misquito, or Spanish. Because the study of knowledge was part of a broader study already in progress, resident researchers knew the linguistic competence of each subject. This knowledge allowed the test to be administered in the language in which the subject felt most comfortable. The test also drew on the help of Tawahka assistants. The use of Tawahka assistants made it easier to administer the test, by allowing the subject to ask for clarification in the language in which she or he felt most comfortable. Female researchers gave the test to women and male researchers gave the test to men. Because the study of folk knowledge came at the end (July 1996) of the longer study (June, 1994-December, 1996), researchers benefited from a level of trust that would have been harder to achieve in an ordinary cross-sectional study.

To minimize the possibility of subjects listening to another person's answers and using that answer as their own, the following steps were taken:

1. all subjects in a village were tested in about three hours by having three researchers give the test at the same time in different parts of the village,

2. the time allowed for a response was limited to 30 seconds for each question,

3. the subjects were tested alone and one of the researchers kept listeners away, and

4. one of the two tests was chosen at random for each subject.

At times, people nearby volunteered answers. When that happened, the question was replaced with a question from the second test and the question was asked when the subject was alone. The time of day that the subject took the test was not recorded. Had it been, the improvement of scores in a village over time could have been estimated, as subjects who heard questions and answers later took the test themselves.

Since subjects had to identify plants and animals from photographs, the test may have been harder for the old and for the illiterate (who had less exposure to printed material) or for those who rely on sound or smell for identification.

Household Socioeconomic and Demographic Surveys

The subject's tests scores were matched to socioeconomic information of the subject collected during the survey the year before (June–August, 1995). Test scores were also matched to the subject's scores in the tests of literacy and arithmetic and to demographic information collected at the start of the panel study in 1994.

The Variables: Definition and Measurement

Table 11-1 contains definition and summary statistics of the variables used in the analysis. The measurement of the variables are defined and explained below.

Dependent Variables: Knowledge of Forest Plants and Animals

Two dependent variables were used: the subject's score in the plant test and the subject's score in the animal test. Answers to questions were marked as correct or as incorrect—there were no open-ended questions. Subjects could score a maximum of eight points in the plant test and seven points in the animal test. The information in table 11-1 suggests that subjects did better in the animal test (mean=3.88; standard deviation=1.05) than in the plant test (mean=2.37; standard deviation=1.46). No subject got a perfect score in either test. Because dependent variables were uncensored, ordinary least square regressions were run, but Huber White robust standard errors were used to correct for non-normal variance in error terms.

TABLE 11-1 Definition and Summary Statistics of Variables

Variable	Definition	Obs	Mean	Sd	Min	Max
Dependent:						
Plantest	Plant test score	80	2.37	1.46	0	7
Animtest	Animal test score	80	3.88	1.05	2	6
Explanatory:						
Male	Sex of subject	80	.52	.50	0	1
Age	Subject's age	80	24.6	16.5	5	70
Education	Maximun education of subject	79	2.06	2.00	0	7
Fluency	Spanish fluency	80	.62	.48	0	1
Literacy	Spanish literacy	80	.31	.46	0	1
Arithmetic	Knowledge of arithmetic	80	.55	.50	0	1
Chickens	# chickens owned	79	10.2	7.52	0	30
Forest	Area of old growth forest cut in tareas (4 tareas=1 ha.)	79	1.46	2.53	0	12
Sharice	% rice harvest sold	77	.11	.14	0	.47
Shafores	% cash income from forest goods sales	77	.09	.18	0	.98
Wageinc	% cash income from wage labor	77	.49	.35	0	1
Krausirpe	Village dummy	80	.45	.50	0	1
Test	Test dummy; test=1 if test I was used (see appendix)	80	.51	.50	0	1

NOTES   The following variables are dummies (name of dummy variable=1): male, fluency in Spanish, literacy in Spanish, math skills, Krausirpe, and test. Shafores and wageinc are share of cash income earned in May, 1995, from sale of forest goods (shafores) or from wage labor (wageinc).

Explanatory Variables Besides Integration to the Market

The tests controlled for the following explanatory variables: age, sex, education, area of old-growth rainforest cleared by household for agriculture in 1995, and wealth (proxied by the number of chickens owned by the household). Tests were given in 1994 to measure literacy in Miskito, Tawahka, and Spanish. Village dummies were also included to control for village fixed effects—a dummy variable was used for the two versions of the test.

Explanatory Variable: Integration to the Market

Three definitions and measures of integration to the market were used:

1. share of rice harvest sold in 1995,

2. share of cash income earned from wage labor in May, 1995, and

3. share of cash income earned from the sale of forest goods in May, 1995.

Different definitions of integration to the market were used because it was hypothesized that integration measured through the first two definitions would erode folk knowledge, but that integration measured through the last definition would enhance it.

In the 1995 socioeconomic and demographic survey, questions on earnings in cash were limited to May (the month before the interview) to enhance the reliability of informant recall. May falls in the rainy season—a time when the Tawahka stay in their villages to farm. Although the choice of May to measure cash income may have produced low estimates, the bias should have affected all households in the same way. Since May is a low point for earning cash, the variation across households may have been less than typically seen, and the variable might therefore have produced weak results.

The test for correlation between explanatory variables found correlation coefficients of less than 0.5, except between Spanish literacy, knowledge of arithmetic, and education. Despite multicollinearity, the variables were left in the regressions because they act as controls and, jointly, affect test scores.

Results

Table 11-2 contains the regression results. The discussion of results is split into three sections. The first section discusses the coefficient of explanatory variables used as controls. The second section discusses the relation between different types of integration to the market and knowledge. The third section contains a discussion of the effects of forest income on knowledge of forest plants and game.

Explanatory Variables: Controls

None of the control variables enhanced or hurt a subject's scores in the animal or plant tests. Men did better than women in the plant test

TABLE 11-2a Socioeconomic Determinants of Indigenous Knowledge

Variable	Integration Through Rice Sale				Wage Labor			
	Plants		Animals		Plants		Animals	
	Coef	Se	Coef	Se	Coef	Se	Coef	Se
Male	.744	.277^2	-.278	.326	.645	.221^3	-.430	.316
Age	.005	.009	.017	.007^2	.007	.008	.015	.007^2
Education	-.169	.096^1	-.039	.095	-.073	.095	-.023	.110
Fluency	.007	.358	.041	.321	-.125	.315	-.164	.332
Literacy	.460	.415	-.140	.429	.074	.363	.030	.433
Arithmetic	.325	.378	.422	.238^1	.472	.350	.473	.245^1
Chickens	.005	.015	.007	.009	.012	.023	.006	.010
Forest	-.041	.053	-.033	.488	-.022	.073	-.039	.042
Krausirpe	-.593	.371	-.324	.292	-.586	.520	-.542	.262^2
Test	-1.039	.412^3	.134	.231	-1.213	.339^3	.006	.219
Constant	3.096	.412	3.596	.428	2.638	.783	4.377	.537
Integration:								
Sharice	-3.423	1.098^3	-.683	.634	n/a	n/a	n/a	n/a
Wageinc	n/a	n/a	n/a	n/a	-.122	.792	-.221	.382
Shafores	n/a	n/a	n/a	n/a	n/a	n/a	n/a	n/a
Obs	76		76		76		76	
R2	41.68		15.36		34.94		16.76	

NOTES Regressions are ordinary least square with Huber White robust standard errors. 1, 2, and 3 significant at £10%, £5%, and £1%. "n/a" means not applicable.

TABLE 11-2b *Socioeconomic Determinants of Indigenous Knowledge*

| | Sale of Forest Goods | | | |
| | Plants | | Animals | |
Variable	Coef	Se	Coef	Se
Male	.677	.238[3]	-.393	.298
Age	.008	.008	.017	.006[3]
Education	-.055	.110	.001	.078
Fluency	-.081	.321	-.125	.333
Literacy	.035	.380	-.022	.420
Arithmetic	.465	.362	.453	.223[1]
Chickens	.016	.026	.011	.011
Forest	-.060	.062	-.073	.035[2]
Krausirpe	-.430	.347	-.357	.303
Test	-.1.67	.351[3]	.055	.206
Constant	2.276	.562	3.477	.447
Integration:				
Sharice	n/a	n/a	n/a	n/a
Wageinc	n/a	n/a	n/a	n/a
Shafores	1.185	.689[1]	1.151	.719
Obs	76		76	
R2	36.63		19.64	

NOTES   Regressions are ordinary least square with Huber
White robust standard errors. 1, 2, and 3 significant at
$\leq 10\%$, $\leq 5\%$, and $\leq 1\%$. "n/a" means not applicable.

(p>|t|=1.3%). Older people, and those who knew arithmetic, did better in the animal test. Those who took Test I did worse in the plant test than those who took Test II (p>|t|=0.5%).

Explanatory Variables: Integration to the Market

Integration to the market—measured by the sale of rice—bore a negative and a statistically significant relation to the score in the plant test (p>|t|=0.5%); the relation between the sale of rice and the score in the animal test was also negative, but statistically insignificant (p>|t|=29.3%). As predicted, people in households that sold a greater share of their rice harvest did worse in both the plant and animal tests. The evidence lends some credence to the idea that integration into the market through the sale of annual crops may be associated with less knowledge of forest plants and game.

Integration into the market through wage labor also appears to be associated with lower scores in the plant and animal tests, but the relations were statistically weak (plants, p>|t|=87.9%; animals, p>|t|=56.9%). Last, integration into the market through the sale of forest goods showed a positive and statistically significant relation with the score on the plant test (p>|t|=9.9%) and a positive, though statistically less significant relation with the score in the animal test (p>|t|=12.3%). As predicted, people from households who drew a greater share of their cash income from the sale of forest goods seemed to know more about forest plants and game.

Specialization

The study tested whether households that drew a greater share of their cash income from the sale of timber and non-timber forest goods knew more about the subset of valuable plants and animals entering commercial channels. Three questions were used in the plant test dealing with plants used for house construction or for making dugout canoes for sale: paleto [*Dialium guianense*], mahogany [*Swietenia macrophylla*], and cortés [*Tabebuia guayacan*]. A dummy variable was created for whether or not subjects answered these questions correctly. Another dummy variable was created for whether or not households answered the question about the social behavior and eating habits of the ocellated antbird (*Phaenostictus mcleannani*) and the great kiskadee (*Pitangus sulphuratus*) correctly (question 14 in the appendix).

The results of probit models (not shown here) with coefficients estimated at the mean value of explanatory variables confirmed, in part, the hypothesis. The share of household cash income from the sale of forest goods bore a positive and statistically significant relation to the score on the animal test (p>|t|=<0.1%) and it bore a positive but statistically weak relation to the score on the plant test (p>|t|=37.3%). People who depend more on the forest for cash seem to know more about the plants and game entering the market.

Conclusion

The results lend some credence to the idea that markets may be associated in systematic ways with the loss and retention of knowledge. Participation in the market for annual crops and (to a lesser extent) for wage labor seems to be associated with erosion of plant and game knowledge. On the other hand, dependence on the sale of timber and non-timber forest goods seems to be associated with higher scores in the plant and animal tests.

Much has been written about the loss of folk knowledge, as once relatively isolated communities increase their economic dependence on the outside world. Many have written about the loss of invaluable knowledge of crops and cultivation practices as rural people switch to modern, high-yielding plant varieties. In agriculture and livestock, trade probably erodes knowledge. It follows that ethno-biological salvage work could have high pay-offs for society.

Things may differ with natural resources. With natural resources, people specialize in extraction—but within a portfolio of game and wild plants. As people become enmeshed in trade, they focus their efforts on some things in the forest at the expense of others. In agriculture and livestock production, trade with the outside world produces replacement—in natural resources, trade may produce a shift in emphasis within an existing suite of forest species. This result has implications for conservation. Although in the long run indigenous people may stop foraging and forget most of what they knew about the forest, loss of knowledge in the immediate future will probably be selective and faster in a subset of species. If this is so and one decides that folk knowledge merits preservation, then efforts to preserve knowledge before it vanishes forever ought to focus on forest plants and game varieties that indigenous people ignore once they enter the world of trade and start to specialize in the world of trade.

Chapter 12

Time Preference, Markets, and the Evolution of Social Inequality

People's valuation of the future affects how much they consume, invest, and save. Because it touches on so many areas, that valuation affects how an entire economy operates. Margo Wilson and her associates point out (Wilson, Daly, and Gordon 1997) that despite the importance of private time preference—the willingness to substitute consumption over time or delay gratification—relatively little is known about its origin or socioeconomic consequences. What little we do know comes mainly from industrial societies (Kirby and Marakovic 1996; Loewenstein 1992; Pender 1996; Thaler and Loewenstein 1989).

Drawing on information from the Tsimane´, this chapter begins to fill the gap by estimating the effect of socioeconomic and demographic variables on people's willingness to delay gratification. The ideas tested come from the writings of social scientists and explanations of patience by the Tsimane´ themselves.

One can explore several new ideas with the information collected. First, the idea that people in simpler rural economies have high rates of discount because they are too poor to wait can be tested. Over half a century ago, Alfred Marshall—one of the founders of modern economics—said that rural people were shortsighted and impulsive. He wrote:

> Whatever be their climate and whatever their ancestry, we find savages living under the dominion of custom and impulse; scarcely ever striking out new lines for themselves; *never forecasting the distant future, and seldom making provision even for the near future; fitful in spite of their servitude to custom, governed by the fancy of the moment....*
>
> (Marshall 1936:723-724 quoted in Yamey 1964:376; author emphasis)

Gary Becker and Casey Mulligan (1997:731-732) reviewed the writings of other researchers who have made the same point as Marshall. This chapter will show that even indigenous forest dwellers with tenuous links to a market economy show variation in their willingness to delay gratification—in fact, most seemed very patient. Second, by turning to a simple economy, one can test whether ideas about time preference developed in Western societies also hold true in different social settings. Third, a study of time preference in a horticultural and foraging economy may reveal patterns that are harder to discern when economies become more complex. The experiment reported in this chapter may be hard to replicate because markets are rapidly absorbing indigenous people in the rainforests of Latin America.

Delay of Gratification Among the Tsimane´

There is no close analog to the concept of time preference in the Tsimane´ language. Time preference or private discount rate refers to a person's willingness to delay gratification. The closest I came to eliciting a concept of time preference was to identify ideas of patience, but patience and time preference are not synonymous. The colloquial word for patience in Tsimane´ (or in English) connotes the idea of delaying gratification, but it also connotes the idea of waiting—even if waiting brings no future gratification. The Tsimane´ word for patience (or to be patient [tyum'chuti or dyichchuti]), connotes the idea of "being quiet," "holding," and "keeping back" for future use.

In some cases, the Tsimane´ do not like to delay gratification. They do not invest in the maintenance of tools and equipment. They throw valuable goods—such as machetes, axes, bicycles, and hand-woven bags (which take many days to make)—in open courtyards. Unlike other lowland Indian groups in Latin America, the Tsimane´ do not plant fruit trees for future use. Their trees are those that sprout on their own. Although this type of behavior may reflect lack of planning, it also reflects rules of ownership. People have clear rights of ownership to land with planted food crops, but they do not have rights of ownership or usufruct to trees (unless they plant trees in their own land). Since any Tsimane´ can use a tree in the forest, they do not have incentive to plant trees in the commons.

Lack of planning for the future can be found at other levels. Wayne Gill, a missionary who has lived and worked with the Tsimane´ for more than two decades translating the Bible, tells the story of offering meat to his linguistic informants many times—generally late at night after Tsimane´ have had dinner. Each time, he has offered the option of keeping the meat for them in his refrigerator until the next day, or allowing them to take the meat home. The Tsimane´ have always taken the meat with them immediately. The decision to take meat with them immediately does not reflect mistrust of missionaries because missionaries and informants have known and worked with each other and lived in the same village for decades.

Gill finds other examples of the lack of ability to delay gratification. He says that he often gives Tsimane´ old magazines as gifts for them to take home:

> Almost invariably they put down the magazine they can look at only in our house but not take home, and begin looking at the gift magazine. There is no thought of: "I will save this for my house."
>
> (Gill March 7, 1997, Letter to the author)

Next to these examples of immediate gratification one can also find examples of waiting for future rewards. The Tsimane´ cut old-growth forest, for example (a hard task given their primitive took kit), to plant annual crops, manioc, and plantains, and wait several months before harvesting the crops. They spend hours (sometimes days) searching for game. Some have started to lobby the municipal government and non-government organizations to increase public expenditures in education so the next generation of Tsimane´ can cope better with opportunities opened by the market. Schooled Tsimane´ have organized themselves to pressure the central government to give them official title to their lands and make it easier to defend themselves against intruders.

During fieldwork, several Tsimane´ were asked to identify and describe the personalities of patient and impatient villagers. Tsimane´ singled out two brothers—Dionisio and Jorge (the names are fictitious), who were roughly 33 and 20 years of age—as examples. Dionisio lives in one of the largest Tsimane´villages, where he finished the fifth grade and excelled as a student.

Many organizations working in the Tsimane´ territory have offered him employment, but he has declined and has decided to stay in his village—hunting and farming. He clears large patches of forest to plant crops for his family and for sale. He is willing to wait many months for crops to mature, and goes hunting for days at a time.

Jorge is more impulsive, the Tsimane´ say, because "he moves around and speaks too much" and likes to work with others. Jorge, like his brother, has studied up to the fifth grade. Unlike Dionisio, Jorge has decided to leave the village and work as a forest guard. At planting time, Jorge rushes to his village from his field station to clear and plant fields, which are small and made in haste. Dionisio says Jorge is impulsive because Jorge has lived and worked in towns and cattle ranches, where he has been exposed to a rapid pace of life.

The Determinants of Time Preferences: Tsimane´ and Western Views

The Tsimane´ have explanations about why some people are more willing to delay gratification than others. The Tsimane´ say that age makes people more willing to wait for future rewards. The young, they say, are more likely to be impulsive—to pack up and move to another village without warning, in search of better game or employment. They also say that willingness to delay gratification runs in families—parents who are willing to wait breed children who are willing to wait, and the Tsimane´ point to examples of the transmission of willingness to wait along kinship lines. Last, the Tsimane´ say women are less likely to wait for gratification than men. They say that women are more likely to want meat now rather than later. Women put pressure on husbands to hunt and grow crops with a short time between planting and harvest.

Drawing on economic and life history theory and empirical studies from Western societies, one can identify other possible correlates of private time preference. Some explanations of outsiders overlap with the explanations of the Tsimane´. The Tsimane´ idea that private time preference runs in families, for example, is echoed by Becker and Mulligan (1997) who say that patient parents make investments to produce patient children.

Becker (1996:11) has hypothesized that education should "improve the appreciation of the future and thereby reduce the discount on the future." Becker and Mulligan (1997) say that investment in education lets people make the future less remote and more vivid, and increase their appreciation of the future:

> ...schooling focuses students' attention on the future. Schooling can communicate images of the situations and difficulties of adult life, which are the future of childhood and adolescence. In addition, through repeated practice at problem solving, schooling helps children learn the art of scenario simulation. Thus, educated people should be more productive at reducing the remoteness of future pleasures. Parents often spend resources on teaching their children to better plan for the future, resources that affect the children's discount rate. Moreover, as children become teenagers and then adults, they experience what had been future utilities, and these experiences also help them to better imagine what the future will be like.
>
> (Becker and Mulligan 1997:735-736)

Becker and Mulligan go on to note that wealth, health, and perhaps income should lower private time preference because they should enhance a person's capacity to imagine the future.

From a life history perspective, anthropologist Allan Rogers (1994) has argued that private discount rates should rise and fall with age. Case studies from the United States have shown a negative relation between private time preference and age (Green, Fry, and Myerson 1994; Thaler 1981; Winston, Woodbury, and Richard 1991). Others have said that people's prospects for longevity affect how much they value the future (Hawkes 1992a; Wilson and Daly 1997; Wilson, Daly, and Gordon 1997). People living in areas with high crime rates may perceive they have a shorter life expectancy and, therefore, act in a shortsighted manner.

Several researchers have shown that income lowers private discount rates because it makes it easier for people to ease borrowing or liquidity constraints, and plan for the future (Cropper, Aydede, and Portney 1992; Hausman 1979; Lawrance 1991). Last, researchers in the United States have found that women have lower discount rates than men (Cropper, Aydede, and Portney 1992; Kirby and Marakovic 1996). In rural India, however, John Pender (1996) found that sex did not affect private discount rates.

Methods and Variables

During June-August, 1996, interviews were conducted with 352 adult (16+ years) Tsimane´ in 209 households and 18 villages, straddling different levels of integration to the market. As discussed in chapter 3, "Research Design," the survey was the second pilot study among the Tsimane´ to examine the effect of markets on their welfare and use of natural resources. Researchers collected psychological, anthropometric, demographic, and socioeconomic information. Each interview lasted 1.5-2 hours.

About 20 minutes into the interview, surveyors asked subjects the following question: "We realize you may be getting tired from answering our questions. We would like to give you a rest. Would you like to have one candy now or two candies at the end of the interview?" If the subject wanted the candy immediately surveyors asked a second question: "Would you like to have one candy now or three candies at the end of the interview?" Depending on the person's response, surveyors delivered the candies on the spot or at the end of the interview. The candies were hard, wrapped in paper, did not melt, and could have been saved for sharing or later consumption. Several mothers said they wanted their candy immediately so they could crack the candy with their teeth and give pieces to their toddlers. Hypothetical questions about time preference (e.g., Barsky et al. 1995) were not asked because they might not have elicited the value people place on the future (Bohm 1994; Cuesta, Carlson, and Lutz 1997) with accuracy.

Based on their responses, subjects were grouped as "very willing," "willing," or "unwilling" to delay gratification. Those subjects labeled "very willing" said from the outset they would wait and take two candies at the end of the interview. "Unwilling" subjects wanted one candy immediately irrespective of the size of later rewards, and "willing" subjects waited until the end of the interview, but only after surveyors had raised the reward from two to three candies. For reasons discussed in the next paragraph, subjects who were "willing" to delay gratification (n=48) were eliminated from the analysis.

The Tsimane´ may have perceived the option of having fewer candies now rather than more candies later as a bargaining game. Those who chose to wait may have been the better bargainers rather than those more willing to wait. Because many hours—and sometimes even several days—were spent in one village, people who had answered questions early could have told other villagers to wait until surveyors raised the rewards. Surveyors did not record the time of each interview, so one cannot determine whether those who took the test later were more likely to turn down the option of having two candies. To remove the potential bias, the "willing" subjects, or those who waited to have three candies at the end of the interview, were removed from the statistical analysis.

Besides time preference, surveyors collected information on the following explanatory variables: total income (wage income + imputed farm income + remittances received), wealth (value of 13 physical assets), illness (days confined to bed the two weeks before the interview), maximum formal education of subject and subject's parents, sex, age, body-mass index, and distance from the village to the town of San Borja.

The Measurement of Time Preference and The Rationale for Using Food

The method for eliciting willingness to delay rewards had several shortcomings that need discussion before presenting the results of the analysis. First, surveyors did not monitor whether or not those who chose the early option ate their candy when they received it or waited until after the interview. If subjects did not eat the candy until later, the early option should not be equated with delayed consumption. Second, since surveyors did the interviews at different times of the day, they may have interviewed some people when they were hungry and others when they were full. Hunger would have affected the preference for candy now rather than later. Surveyors tried to proxy for hunger by measuring body-mass index. Third, some Tsimane´ might have mistrusted the researchers and felt that their chances of getting the delayed option were lower. Taking a candy now would have eliminated the trust problem. Last, some mothers opted to have the candy immediately so that they could give it to their children and reduce the distractions in answering questions during the interview.

We chose candy rather than money, clothing, or other goods to measure the ability to delay gratification for three reasons. First, the rewards had to be carried to villages—generally on foot and often far away. It was difficult to take bulky or heavy goods. Second, the use of other light, portable items, such as cigarettes or coca leaves was unethical. Third, the Tsimane´ like candy. They often buy it for their families when they go to town and give it to each other as gifts.

The choice of food to measure the ability to delay gratification fits with the findings of social psychologists that people's preference for food is strong and mirrors their ability to delay gratification or make impulsive choices (Kirby and Marakovic 1996; Mischel, Shoda, and Rodríguez 1989). The choice of candy to measure private time preference over only 1-2 hours might not capture with accuracy time preference for investments, which take place over a longer time.

Information and Econometric Model

Table 12-1 contains definitions and descriptive statistics of the variables used in the analysis. Although surveyors interviewed 352 subjects, information from only 257 subjects could be used. The loss of 95 subjects comes from eliminating the "willing" category and from missing values for the variable age. Many Tsimane´ could not estimate their age and surveyors did not attempt to place them in an age cohort.

TABLE 12-1 Summary Statistics of Variables

Variable	Mean or Percent	Sd	Min	Max
Dummies:				
Very willing	89.88%			
Female	35.01%			
Parental education	14.39			
Continuous:				
Age	30.66	11.56	16	78
Education	1.74	2.40	0	14
BMI	23.03	2.54	17	14
Income	2368	2464	11	14712
Wealth	2912	2729	110	17543
Illness	3.04	3.77	0	14
Distance	29.89	15.31	3	82

NOTES For dummy variables percent rather than means reported. Name of dummy variable=1. For meaning of "very willing" see text; excluded category is "unwilling". Continuous variables: age=age of subject in years; education=years of completed education; BMI=body-mass index (kg/mt^2); wealth=value in *Bolivianos* of 13 assets (e.g., animals, tools); illness=days confined to bed during two weeks before the interview; income=imputed farm income+wage income+remittances in *Bolivianos* (1 US=5.05 *Bol*); distance=distance in kilometers in a straight line from village to town of San Borja using a geographic positioning system receiver. n=257.

Close to 90 percent of the 257 subjects were "very willing" to delay grat-ification. The mean age of subjects in the sample was 30 years. Thirty-five percent of the subjects were women. Only 14.39 percent of the sample had a parent who had attended school. The average subject had 1.74 years of schooling. During the two weeks before the interview, subjects reported hav-ing been ill 3.04 days. Average household income and wealth in 1995 were 2,368 and 2,912 *Bolivianos*—both variables displayed high variance (1 U.S. dollar=5.05 *Bolivianos*). The average village was 29.89 km in a straight line from the town of San Borja.

A multivariate probit model is used to estimate the effect of explanatory variables on the probabilities of being "very willing" to delay gratification. As in other probit estimations in this book, probabilities are estimated at the mean value of explanatory variables. A probit model was chosen over dis-criminant analysis because discriminant analysis works less well when using dummies as explanatory variables (Kennedy 1993:236; Press and Wilson 1978). In the analysis, dummies are used for variables such as sex and parental education. Since probit and logit models produce similar results when estimating probabilities at the mean value of explanatory variables (as done here), the choice of a probit regression over a logit regression is arbi-trary and does not affect the results. The regression of table 12-2 was re-esti-mated using a logit model and yielded essentially the same results.

Results

Table 12-2 contains the regression results. The results suggest that school-ing is associated with a reduced willingness to delay gratification. One more year of schooling decreased the probability of a subject being "very willing" to wait by 1.69 percent (z=-2.41; p>|z|=1.6%). The only other variable that showed a statistically significant relation, at the 95 percent confidence level or above, was illness. Each additional day of illness above the mean of 3.04 days from the sample used in the probit regression increased the likelihood of being "very willing" to wait by 0.98 percent (z=2.04; p>|z|=4.2%).

The income variable bore the correct, positive sign predicted by econo-mists, but the wealth variable did not (Cropper, Aydede, and Portney 1992; Hausman 1979; Lawrance 1991)—both variables were statistically insignifi-cant. Contrary to what the Tsimane´ say, women seemed to be more willing to wait than men, and age did not seem to make people more willing to

TABLE 12-2 *Covariates of Being "Very Willing" to Delay Gratification*

Variable	Coef	Robust Se	Z	p>|z|	Mean of X variable
Age	-.02	.17	-.15	.88	30.66
Female	3.71	3.62	.91	.36	.35
Education	-1.69	.71	-2.41	.01	1.74
Parental education	-.87	4.70	-.19	.84	.14
BMI	-.20	.56	-.36	.72	23.03
Income	1.93	1.60	1.18	.23	7.20
Wealth	-2.45	2.12	-1.17	.24	7.59
Illness	.98	.52	2.04	.04	3.04
Distance	-.12	.11	-1.16	.24	29.89
Pseudo R2	.08				
Obs	257				

NOTES Regression is probit with probabilities estimated at the mean value of explanatory variables. Probit includes robust standard errors and no constant. Coefficients are probabilities. Dependent variable is dummy, "very willing" to delay gratification (1=very willing, 0=unwilling; "willing" category deleted, see text). Income and wealth in logarithms.

delay gratification. Being a woman was associated with a 3.71 percent greater likelihood of waiting for future rewards (z=0.91; p>|z|=36%). People from remote villages were less willing to wait. None of the findings discussed in this paragraph were statistically significant at the 90 percent confidence level or above.

A Hypothesis About Time Preference and Occupational Choice

This section compares the mean of socioeconomic and demographic variables between two samples—unwilling and very willing to delay gratification—to explore the link between occupational choice and time preference. Table 12-3 contains the results of the comparison. In table 12-3, the "very willing" group has been added back to the "willing" group and both are called "very willing." This was done to increase the sample size. The results of the analysis discussed in this section do not change if one uses the smaller sample.

TABLE 12-3 Covariates of Willingness to Delay Gratification

Variable	Very Willing (n=258; 90%)			Unwilling (n=29; 10%			Equality of Means Test	
	Obs	Mean	Sd	Obs	Mean	Sd	t	p>|t|
Income (Bol):								
Total	248	2229	146	27	2499	600	.55	57.88%
Farm	248	1131	70	27	827	212	-1.35	17.80%
Wage	250	1067	139	29	1815	630	1.62	10.58%
Wealth (Bo):	250	2701	148	29	3437	718	1.47	14.12%
Forest (ha):								
Old growth	250	.73	.05	29	.57	.08	-.95	34.00%
Secondary	250	.74	.05	29	.50	.07	-1.36	17.45%
House size:	250	4.87	.19	29	4.10	.43	-1.32	18.62%

NOTES   "Very willing" includes original "very willing" and "willing" groups (see text). Farm and wage refers to farm and wage income in *Bolivianos* ($1US = 5.05Bol). Forest refers to area of old and secondary-growth forest cleared. Household size is number of people in the household.

The information in table 12-3 suggests that people who are very willing and unwilling to delay gratification specialize in different occupations. People very willing to wait seem to earn more income from the farm, and those unwilling to wait seem to earn more income from wage labor. The occupational choice may explain some of the socioeconomic and demographic differences between the two populations.

People unwilling to delay gratification earned higher income from wage labor than those very willing to wait. Impatient subjects earned 1,815 *Bolivianos* a year from wage labor, whereas patient subjects only earned 1,067 *Bolivianos*. An unmatched *t* test, comparing the equality of means, shows that the difference was statistically significant (p>|t|=10.58%). Since people unwilling to wait work more in wage labor, they spend more time in towns and in logging camps away from the village. From this fact, several findings fall into place.

Because they have a toehold in the modern world, people unwilling to delay gratification may not be able to spend as much time in the village farming or foraging. They clear smaller plots of forest. People unwilling to wait to clear half a hectare of either old or secondary-growth forest, whereas people very willing to wait to clear about three-quarters of a hectare. Using money rather than candy among a much larger and different sample of Bolivian Amerindians produced the same results. Table 5-5 shows that people with higher private discount rates cut less old-growth forest, although the effect was physically small and statistically insignificant. Among the Tsimane´, people unwilling to wait earned 827 Bolivianos a year from farming, compared with 1,131 Bolivianos earned by people very willing to wait. Perhaps people unwilling to wait have a smaller household size (4.10) than subjects very willing to wait (4.87) because impatient people are linked to the urban labor market. With a foothold in the modern world and higher total income, people unwilling to wait find it easier to amass wealth. People unwilling to wait had higher wealth (3,437 Bolivianos) than people very willing to wait (2,701 Bolivianos).

The information in table 12.3 suggests that people willing to delay gratification seem to stay in the village. Anchored to the countryside, patient people naturally cut more forest, earn higher incomes from the farm but lower incomes from wage labor, and have less wealth but bigger families. Although none of the results discussed in this section are statistically significant at the 90 percent confidence level or above, they point to a pattern that could become significant with a larger sample.

To estimate the extent to which willingness to delay gratification affects a person's earnings from wage labor, a Mincer-like earnings function (not shown) was run with the logarithm of wage income as the dependent variable and the following explanatory variables: age, experience (defined as age minus education minus five), experience squared, education, sex, household size, dummies for villages, and the variable for willingness to delay gratification. The coefficient on the willingness-to-wait variable bore the correct sign. People willing to wait earned three times less from wage labor than those unwilling to wait, but results were statistically weak ($p > |t| = 41\%$).

Conclusion

Three tentative conclusions emerge from the analysis. First, contrary to what one might have expected from people in a horticultural and foraging society, most (89.88%) Tsimane´ were willing to delay gratification. Even in this simple rural economy, one finds variation in the willingness to delay gratification. Second, contrary to what Becker and Mulligan say (Becker

1996:11; Becker and Mulligan 1997), schooling seems to make people less willing to delay gratification. Last, willingness to delay gratification seems to be associated with occupational choices—which, in turn, correlate with demographic and socioeconomic attributes of the subject.

As rural economies modernize, the drift to different occupations has implications for the rise of social inequalities. People unwilling to delay gratification seem to end up in wage labor and accumulate more wealth than their patient relatives in the forest. Only further empirical work with better models, larger samples, and with a better metric of time preference will allow us to decide if the idea holds up.

Part III

WHAT WE HAVE LEARNED

Because markets have so often grown alongside conquest, migration, and cultural change, people have attributed the effects of other processes to markets. Like the birth of most new institutions, markets, when first introduced, caused stress. Markets undermined the role of traditional healers and craftsmen. Once they acquired guns, metal pots, and plastic buckets, indigenous people enclosed their huts or moved away from communal houses to hide their wealth. Soon after contact, Indians in North and South America—in parallel fashion—started to fight with each other to get Western goods (Chagnon 1992:260-261; Ferguson 1992, 1995; Rich 1960). Soon after contact, many lowland groups, including the ones discussed in this book, retreated farther into the forest to avoid enslavement by Westerners and indigenous groups acting on behalf of Westerners.

But markets have also enlarged "the world of ideas, of technology, and of possibilities" unimagined by the ancestors of indigenous people today (Sowell 1998:5). Even without coercion, indigenous people drifted to the market to exchange pelts, canoes, and food for guns, steel tools, cloth, and a wide range of luxuries (such as "tobacco, spirits, gay cloth of different kinds, beads, and caps") (Rich 1960:45). Steel tools, guns, and cloth displaced traditional technologies for hunting and manufacturing because they allowed indigenous people to produce with less effort.

Because markets have provided indigenous people with so many improved goods, indigenous people have been reluctant to return to their old ways once they enter the market. Kim Hill and Magdalena Hurtado say that after many epidemics, the Aché of Paraguay (one of the last full-time foragers in Latin America):

> ...have not had a difficult time adjusting to their new lifestyle, and they do not particularly regret the loss of many of their old cultural patterns. None of them expresses a desire to return to their forest lifestyle despite some naive attempts by well-meaning anthropologists and indigenous rights workers to convince them of the desirability of their former way of life...Their biggest dreams are usually to acquire a shotgun, some new clothes, and perhaps a radio, and to have many healthy children grow to adulthood.
> (Hill and Hurtado 1996:78)

Tallying all the positive and negative effects of markets on the culture and landscape of lowland indigenous people is hard because the effects of markets appear in both direct and indirect ways—in the present and in the distant future. Rather than provide a complete tally of all the effects, this book presents a new way of thinking about how to assess the effect of markets on the quality of life and habitat of indigenous people. Using quantitative information from several lowland cultures, it has tried to find regularities in the process of integration to the market. The results of the comparison yield several lessons in theory and methods—for both anthropologists and policymakers. These lessons are detailed in this final chapter of the text.

Chapter 13

CONCLUSIONS

This chapter contains a summary of the book's contribution to anthropological theory, anthropological methods, and public policy.

Contribution to Anthropological Theory

There is no unified theory to explaining in a parsimonious way what happens to the habitat, society, ideas, quality of life, and material culture of indigenous people as they become part of the market. Rather than develop such a theory or wait until someone else does, this book has presented the results of empirical explorations before the indigenous people we know today disappear forever. Commercial goods and relations have seeped into lowland Indian societies for centuries, but the rate of exposure to the outside world and the rate of cultural change have increased.

This book has relied on trade and on price theory to explain, in a serial way, the effect of markets on the life and habitat of indigenous people experiencing greater integration into a market economy. At times, as happened with the analysis of reciprocity, demography, and leisure, the received wisdom from anthropology and sociology was used and tested to see how well it did against the facts. At other times, as happened with the analysis of health, past ethnographies were used to formulate and test hypotheses. On new topics—such as the origins of private time preference—this work drew on the best thinking from the social sciences to see how well theory and empirical findings from Western societies did in more exotic settings.

The largest empirical lesson from the analysis is that markets seem to produce unclear, sometimes benign, and sometimes harmful effects on quality of life and the environment. Trade and price theory predict some of the ambiguity. Trade and price theory suggest that markets should produce unclear effects on the loss or retention of plant and animal knowledge or on the sustainability of natural resources extracted by indigenous people. Price theory suggests that markets should both increase and decrease the amount of tropical rainforest cut by households, depending on the level of income of households and on how households integrate to the market. Price theory also predicts that markets may increase or decrease the amount of leisure available to people, depending on the relative strength of the income and the substitution effect. Some of the ambiguity found in the empirical analysis, however, may have other roots. Mismeasurement of variables could have either biased coefficients toward zero, increased their standard error, or decreased their level of statistical significance, leading us to the false conclusion that markets had a small or negligible effect on the observed outcome.

But what contributions have been made to some of the larger topics mentioned in chapter 1, "The Question and its Significance," such as the evolution of social solidarity, the evolution of inequality, or the role of infrastructure or superstructure in behavior?

Hopefully, this book has presented evidence to question the adequacy of safety nets in autarky and their weakening with the onslaught of markets. The finding is surprising because anthropologists routinely portray lowland Indians as people enmeshed in thick webs of reciprocal exchanges. Though lowland Amerindians may exchange goods and services with each other, the people studied did not seem to help their less fortunate brethren in times of need—or at least did not seem to help them as much as many people seem to think. Forty-three of the subjects who fell ill during the forest-clearing season had to weather the episode alone—with no help in cash, goods, or in services from their neighbors or their family.

On the rise of inequality, we have shown—in a frankly exploratory way—that the seeds of socioeconomic differentiation may already lie dormant in simple economies. Even people in a simple horticultural and foraging economy show variation in their willingness to delay gratification. People with different private time preference seem to differ in their use of natural resources, in their level of wealth, in demographic attributes, and in the occupations they pursue. Impatient people seem to clump in modern occupations and have more income from wage labor than patient people—who seem to gravitate to rural jobs. Whether or not wage labor allows people to become patient or whether patience allows people to work for wages and become part of the modern economy still needs to be settled with further empirical work.

Last, a multivariate approach was used to test the weight of infrastructure, structure, and superstructure, and, in so doing, move the debate in anthropology beyond its present stalemate. Most of the econometric models used in the book contain a mix of socioeconomic, demographic, and cultural explanatory variables. In chapter 9, "Human Health: Does it Worsen with Markets?", the statistical weight of variables related to the market was compared with variables related to acculturation in shaping morbidity. The analysis showed that both types of variables mattered in a statistical sense. Chapter 5, "Forest Clearance: Income, Technology, and Private Time Preference," tested whether a cognitive variable—private time preference—affected the decision to clear forest, and found that it exerted a small effect.

Contribution to Anthropological Methods

The most important contribution of this book does not lie in the empirical findings. Most empirical findings in the social sciences do not survive the test of time, and the findings reported in this book are no exception. The more lasting contribution of this book hopefully lies in a new, more rigorous way of studying an old topic.

Looking back on the methods used to collect information, one can identify several things which one should do differently to strengthen future research. First, it would have been beneficial to have sharper models before collecting information. The low R squares in so many of the regressions shows that researchers still do not understand well what happens to outcomes such as health or private time preference as people become part of the market. Better models should help future researchers to collect more relevant information.

Second, a better method would have combined aspects of a panel and a cross-sectional study. A baseline study among several scattered groups could have been done and researchers could have returned every few years to measure changes. To do continuous research in a village for many years costs too much. Large socioeconomic and biological changes within the same person, household, or habitat generally do not become prominent until many years have elapsed, so measurements in contiguous years often do not yield as much variation as measurements over a longer time. A longer panel would have also allowed a better understanding of the effects of integration to the market on indigenous people as the process unfolded.

Third, one would have benefited from larger samples of ethnic groups and communities to increase trust in the empirical results. Many ethnic groups are needed to ensure that the results are not unique to one and only one group. Many communities are needed because behavior reflects the interaction of community with personal variables. The use of contiguous cultures may not have provided enough variance.

Last, except for experimental studies, there are few ways to control well for biases from potential reverse causality. In analyzing forest clearance, illness was used as an instrumental variable to control for biases from possible endogeneity. In the chapters on folk knowledge and leisure, some of the more endogenous explanatory variables were lagged. For private time preference, however, causality could have run both ways—from unwillingness to delay gratification to greater wage income, or from greater wage income to unwillingness to delay gratification. As discussed in chapter 1, "The Question and its Significance," giving money to indigenous people selected at random may be the only way of controlling for endogeneity and measuring the pure effect of markets on social and biological outcomes. Researchers have yet to take that bold step.

Knowledge and Public Policy

People fall into three camps when discussing the role government should play with lowland indigenous populations. Some would like the rest of the world to leave indigenous people alone to minimize the effect of markets. Though more prevalent in the past, the isolationist position has waned because the public realizes the inevitability of market expansion and the high cost of keeping indigenous people in near-autarky. Others would like to see the *status quo* prevail, allowing markets, encroachers, and cultural contact to spread unfettered by regulations. A third group would like to see indigenous groups enjoy more autonomy in deciding their future. They

would like to see government level the playing field through public investment and laws safeguarding the property rights of indigenous people.

Most policy-makers in Latin American fall into the second camp. Though
more and more policy-makers honor the autonomy of indigenous groups in
the abstract, most of them do not protect the rights of indigenous people
with vigor or lobby for larger or more efficient public investment in indigenous territories. Several reasons may explain why policy makers in Latin
America take a hands-off attitude toward the fate of lowland Indians.

The first reason has to do with the lack of political will. Top policy-makers look the other way when encroachers move into indigenous territories
because indigenous people in the lowlands—with few exceptions—carry little political weight in national politics. Few and scattered, lowland Indians
lack lobbying pressure to bring about changes in the nation's laws that matter most—such as legislation on land titling.

But, even when they have the political will, top Latin American policy-
makers do not know where to invest or what public policies to put in place
to improve the quality of life of indigenous people and conservation. These
are topics in which there has been little applied research. True, cultural
anthropologists and conservation biologists have been studying lowland
groups and natural resources for decades, but little of that research focuses
on policy. That research contains much background material and excellent
diagnosis, but it does not tell public officials what to do or how to do it in a
fiscally responsible and politically practical way.

One cannot turn to leading international development organizations for
current information and new ideas because they, too, have ignored primary
research on public policy among lowland Indians. Leading development
organizations do virtually no primary research among lowland Indians, and
rely on outdated, secondary information—most of which is unsuited for the
design of public policies. More interested in pushing money through safe
projects with visibility and doing what the Washington consensus demands
than obtaining better information to help governments formulate policies,
large international bureaucracies lack vision of where to go with indigenous
people. The lack of interest in collecting primary information and the poverty of the policy research they do on lowland Indians contrasts with the
large amount of research they do on topics such as macroeconomic management, tax administration, or trade reform.

Latin American governments and international institutions need to take
the collection of primary information in policy research for lowland Indians

more seriously because many of the questions central to the welfare of lowland populations and conservation lack clear answers and require the collection of primary information—at present unavailable. A few examples will clarify the point.

Weak states, such as those found in most of Latin America, find it hard to protect human rights or deliver public services to indigenous people who live in distant lands. Weak states have little money and poorly trained public servants. Policy-makers who wish to carry out reforms on behalf of indigenous people would have a hard time designing policies or putting them in place because of administrative and financial constraints.

Consider the delivery of public services in indigenous territories. Many indigenous territories in the lowlands of Latin America are swarming with non-government organizations and church groups that replicate each other's work and do not know or care to know what their neighbors do. Policy-makers in Latin America do not know who does what or where in lowland indigenous territories. It is not clear what the government should do to coordinate or make the best use of the talent, financial resources, and capital goods that these institutions bring. Weak states breed institutional chaos in the hinterlands. Ironically, the presence of so many organizations in the countryside makes it easier for policy-makers to justify withdrawing government support. Why should the government invest in lowland indigenous populations when so many other institutions are already working in the area?

Consider the problem of land titling. A policy-maker would not know whether to grant titles to households, communities, ethnic groups, or to a federation made up of many ethnic groups. There are many private opinions and political agendas on the subject, but few empirical studies exist that assess what would work best for lowland Indians and the nation.

Gaps in basic and applied research make it necessary for Latin American governments and international development organizations to take the role of empirical, public-policy research in the lowlands more seriously. Until officials in government and development organizations have more information and a better understanding of lowland populations, they will find it hard to put public policies in place that bring about progressive changes. Knowledge, of course, will not substitute for the lack of political will. Both will need to go hand in hand.

Appendix

Test of Folk Knowledge

Each test contained two sections, one for plants and one for animals. In the first four questions of each section, subjects had to identify plants or animals. To identify plants and animals, we showed subjects pictures of birds, mammals, plants, or plant parts from the books by Ridgely and Gwynne (1989), Emmons (1990), and Thirakul (n.d.). Our edition of the book by Thirakul has no date of publication. Numbers after "n.d." below refer to page numbers. In the second section of each test, subjects had to answer questions about the ecology of plants or the habits of animals.

Plants. The first four questions concern identification. After question 4, correct answers are shown in brackets.

Test I	Test II
1. Mahogany (*Swietenia macrophylla*) (Thirakul n.d.:266)	1. Paleto (*Dialium guianense)* (Thirakul n.d.:154)
2. San Juan (*Vochysia hondurensis*) (Thirakul n.d.:458)	2. Guácimo (*Luehea seemannii*) (Thirakul n.d.:436)
3. Laurel (*Cordia alliodora*) (Thirakul n.d.:132)	3. Kerosín (*Tetragastris panamensis*). (Thirakul n.d.: 146)
4. Cedro (*Cedrela odorata*) (Thirakul n.d.:258)	4. Masika (*Brosimum alicastrum*) (Thirakul n.d.:290)
5. Does mahogany (*Swietenia macrophylla*) fruit in the dry season? [yes] (Thirakul n.d.:266)	5. Is the fruit of suita macho (*Geonoma sp.*) white? [no] (No picture used)
6. Does cortés (*Tabebuia guayacan*) flower in the dry season? [no] (Thirakul n.d.:112)	6. Does paleto (*Dialium guianense*) have trunk buttresses (*gambas*)? [yes] (Thirakul n.d.:154)
7. Does guapinol (*Hymenaea courbaril*) have milk? [yes] (Thirakul n.d.:156)	7. Does jobo (*Spondias mombin*) lose all its leaves in the dry season? [yes] (Thirakul n.d.:78)
8. Does tambor (*Schizolobium parahybum*) have broad leaves? [no] (Thirakul n.d.:158)	8. Does naranjo (*Terminalia amazonia*) have trunk buttresses? [yes] (Thirakul n.d.:174)

Animals. In questions 9-12 subjects were asked if the animal was found in the Tawahka territory.

Test I	Test II
9. Blue-crowned motmot (*Momotus momota conexus*) (Ridgely and Gwynne 1989: plate 16, no 3). [yes]	9. Lance-tailed Manakin (*Chiroxiphia lanceolata*) (Ridgely and Gwynne 1989: plate 27, no. 3). [no]
10. Tayra (*Eira barbara*) (Emmons 1990: plate 16, no 5). [yes]	10. Mantled howler monkey (*Alouatta palliata)* (Emmons 1990: plate 13, no. 5). [yes]
11.Spot-crowned barbet (*Capito m. maculicoronatus)* (Ridgely and Gwynne 1989: plate 17, no. 4). [no]	11. Baird's Trogon (*Trogon bairdii)* (Ridgely and Gwynne 1989: plate 15, no. 9). [no]
12. Masked Tityra (*Tityra semifasciata*). (Ridgely and Gwynne 1989; plate 26, no. 14) [yes]	12. Spotted Antbird (*Hylophylax n. naevioides*)(Ridgely and Gwynne 1989: plate 22, no. 5). [yes]
13. Does the pava (*Penelope purpurascens*) live high in the trees? (Ridgely and Gwynne 1989) [yes]. (Picture of pava was not a plate, but a black and white illustration on p.114.)	13. Does the armadillo *(Dasypus novemcintus)* live in the trees? (Emmons 1990: plate a, no.3). [no].
14. Does the ocellated antbird (*Phaenostictus mcleannani*) occur with ants? (Ridgely and Gwynne 1989: plate 22, no. 6). [yes].	14. Does the great kiskadee (*Pitangus sulphuratus*) eat insects? (Ridgely and Gwynne 1989: plate 23, no.15) [yes].
15. Does the blue-crowned motmot (*Momotus momota conexus*) nest in trees? (Ridgely and Gwynne 1989: plate 16, no.3). [no]	15. Does the buff-throated saltator (*Saltator maximus*) live in old-growth forest? (Ridgely and Gwynne 1989: plate 37, no. 9). [no].

References

Allen, J. C. and D. F. Barnes. 1985. The causes of deforestation in developing countries. *Annals of the Association of American Geographers* 75:2:163-184.

Alston, L. J., G. D. Libecap, and R. Schneider. 1996. The determinants and impact of property rights: Land titles on the Brazilian frontier. *The Journal of Law, Economics & Organization* 12:1:25-61.

Alvarado, M. 1996. Uso del monte alto en comunidades indígenas del Beni: el efecto del mercado en el tamaño del chaco. B. A. thesis, Universidad Nacional Mayor de San Andrés, La Paz, Bolivia.

Amin, R. 1997. NGO-promoted women's credit program. *Women & Health* 25:1:71-88.

Anderson, A. B., P. H. May, and M. Balick. 1991. *The Subsidy from Nature: Palm Forests, Peasantry and Development on an Amazon frontier.* New York: Columbia University Press.

Añez, J. 1992. The Chimane experience in selling jatata. In M. J. Plotkin and L. M. Famolare, eds., *Sustainable Harvest and Marketing of Rain Forest Products*, pp. 197-198. Washington, D.C.: Island Press.

Appadurai, A. 1986. Introduction: Commodities and the politics of value. In A. Appadurai, ed., *The Social Life of Things*, pp. 3-63. Cambridge: Cambridge University Press.

Arrow, K., B. Bolin, R. Costanza, P. Dasgupta, C. Folke, C. S. Holling, B. O. Jansson, S. Leven, K. G. Meller, C. Perrings, and D. Pimentel. 1995. Economic growth, carrying capacity, and the environment. *Science* 268:520-1.

Aspelin, P. 1979. Food distribution and social bonding among the Mamainde of Mato Grosso, Brazil. *Journal of Anthropological Research* 35:309-327.

Audretsch, D. B. and M. P. Feldman. 1996. R & D sillovers and the geography of innovation and production. *American Economic Review* 86:3:630-41.

Bailey, K. V. and A. Ferro-Luzzi. 1995. Use of body-mass index of adults in assessing individual and community nutritional status. *Bulletin of the World Health Organization* 73:5:673-680.

Baker, P., J. Hanna, and T. Baker. 1986. *The Changing Samoans. Behavior and Health in Transition*. New York: Oxford University Press.

Barsky, R. B., M. S. Kimball, F. T. Juster, M. D. Shapiro. 1995. Preference parameters and behavioral heterogeneity: An experimental approach in the health and retirement survey. Cambridge, Mass.: National Bureau of Economic Research, Working Paper No. 5213.

Bautista, R. M. and A. Valdés, eds. 1993. *The Bias Against Agriculture: Trade and Macroeconomic Policies in Developing Countries*. San Francisco: ICS Press for the International Food Policy Research Institute.

Bawa, K. S. and S. Dayanandan. 1997. Socio-economic factors and tropical deforestation. *Nature* 386: 562-3.

Becker, G. S. 1965. A theory of the allocation of time. *The Economic Journal* 75:299:493-517

Becker, G. S. 1996. *Accounting for Tastes*. Cambridge, Mass.: Harvard University Press.

Becker, G. S. and C. B. Mulligan 1997. The endogenous determination of time preference. *Quarterly Journal of Economics* 112:3:729-759.

Bedoya, E. 1991. *Las causas de la deforestación en la amazonia peruana: un problema estructural*. Lima: Centro de Investigaciones y Promoción Amazónico.

———. 1995. The social and economic causes of deforestation in the Peruvian Amazon basin: Natives and colonists. In M. Painter and W. H. Durham, eds., *The Social Causes of Environmental Destruction in Latin America*, pp. 217-248. Ann Arbor, Michigan: University of Michigan Press.

Behrens, C. A. 1986. The cultural ecology of dietary change accompanying changing activity patterns among the Shipibo. *Human Ecology* 14:4:367-396.

———. 1990. Qualitative and quantitative approaches to the analysis of anthropological data: A new synthesis. *Journal of Quantitative Anthropology* 2:305-328.

———. 1992a. A formal justification for the application of GIS to the culture ecological analysis of land use intensification and deforestation in the Amazon. Paper presented at a conference on "Anthropological Applications of Human Behavior Through Geographical Information and Analysis." University of California, Santa Barbara, National Center for Geological Information and Analysis. February 1st and 2nd, 1992.

——. 1992b. Labor specialization and the formation of markets for food in a Shipibo subsistence economy. *Human Ecology* 20:4:435-462.

Behrman, J. R. and A. B. Deolalikar. 1987. Will developing country nutrition improve with income? A case study for rural south India. *Journal of Political Economy* 95:3:492-507.

——. 1993. Unobserved household and community heterogeneity and the labor market impact of schooling: A case study for Indonesia. *Economic Development and Cultural Change* 41:3:461-488.

Bell, C., P. Hazell, and R. Slade. 1982. *Project Appraisal in a Regional Perspective.* Baltimore, Maryland: The Johns Hopkins University Press.

Bellon, M. R. and J. E. Taylor. 1993. 'Folk' soil taxonomy and the partial adoption of new seed varieties. *Economic Development and Cultural Change* 41:4:763-787.

Benjamin, D. 1992. Household composition, labor markets, and labor demand: Testing for separation in agricultural household models. *Econometrica* 60: 2: 287-322.

Berry, J. W., U. Kim, T. Minde, and D. Mok. 1987. Comparative studies of acculturative stress. *International Migration Review* 21:491-511.

Bliege Bird, R. L. and D. W. Bird. 1997. Delayed reciprocity and tolerated theft. *Current Anthropology* 38:1:49-78.

Bodley, J. H., ed. 1988. *Tribal Peoples & Development Issues: A Global Overview.* Mountain View, California: Mayfield Publishing Company.

Bohm, P. 1994. Time preference and preference reversal among experienced subjects: The effects of real payments. *Economic Journal* 104:427:1370-1379.

Bohn, H. and R. T. Deacon. 1996. Ownership risk, investment, and the use of natural resources. *American Economic Review* 90:526-549.

Boster, J. S. 1984. Inferring decision making from preferences and behavior: An analysis of Aguaruna Jívaro manioc selection. *Human Ecology* 12:343-358.

——. 1986. Exchange of varieties and information between Aguaruna manioc cultivators. *American Anthropologists* 88:2:428-436.

Boya, J. 1996. Learning by doing and the choice of technology. *Econometrica* 64:6:1299-1311.

Brinkman, U. K. 1994. Economic development and tropical disease. *Annals of the New York Academy of Sciences* 740:303-311.

Brown, P. J. and E. D. Whitaker. 1994. Health implications of modern agricultural transformations: Malaria and pellagra in Italy. *Human Organization* 53:4:346-351.

Browner, C. H. 1991. Gender politics in the distribution of therapeutic herbal knowledge. *Medical Anthropology Quarterly* 5:99-132.

Bruhn, J. G. and S. Wolf. 1979. *The Roseto Story: An Anatomy of Health*. Norman, Oklahoma: University of Oklahoma Press.

Brush, S. B. 1993. Indigenous knowledge of biological resources and intellectual property rights: The role of anthropology. *American Anthropologist* 95:3:653-686.

Brush, S. B., J. E. Taylor, and M. R. Bellon. 1992. Technological adoption and biological diversity in Andean potato agriculture. *Journal of Development Economics* 39:2:365-388.

Bunker, S. 1985. *Underdeveloping the Amazon*. Urbana-Champaign, Illinois: University of Illinois Press.

Burkhalter, S. B. and R. F. Murphy. 1989. Tappers and sappers: Rubber, gold, and money among the Mundurucú. *American Ethnologist* 16:1:100-116.

Cahn, P. S. 1996. To See the Forest for the Trees: Changing Perceptions of the Rain Forest Among Tawahka Amerindians. B. A. thesis, Harvard University.

——. 1996a. Las historias de los Tawahka. Cambridge, Mass.: Department of Anthropology, Harvard University. Manuscript.

Caicedo, M. D. 1993. Wildlands Conservation and Community Resource Management: A Critical Analysis of the Proposed Tawahka Biosphere Reserve, La Mosquitia, Honduras. M. A. thesis, San Jose State University.

Camerer, C., G. Lowenstein, and M. Weber. 1989. The curse of knowledge in economic settings: An experimental analysis. *Journal of Political Economy* 97:5:1232-1255.

Carneiro, R. L. 1983. The cultivation of manioc among the Kuikuru of the Upper Xingu. In R. B. Hames and W. T. Vickers, eds., *Adaptive Responses of Native Amazonians*, pp. 65-112. New York: Academic Press.

Carrier, J. 1990. Gifts in a world of commodities. *Social Analysis* 29: 19-37.

Carter, M. R. 1997. Environment, technology, and the social articulation of risk in West African agriculture. *Economic Development and Cultural Change* 45:3:557-590.

Cashdan, E. 1985. Coping with risk: Reciprocity among the Basarwa of northern Botswana. *Man* 20:454-74.

———. 1987. Trade and its origins on the Botletli River, Bostwana. *Journal of Anthropological Research* 43:2:121-138.

Castillo, F. 1988. *Chimanes, cambas y collas*. La Paz: Don Bosco.

Censo Indígena. 1994-1995. *Censo Indígena del Oriente, Chaco y Amazonia*. La Paz: Secretaria de Asuntos Etnicos, de Genero y Generacionales, Ministry of Human Development, Government of Bolivia.

Chagnon, N. A. 1992. *Yanomamö. The Last Days of Eden*. San Diego, California: Harcourt Brace & Company.

Chayanov, A. V. 1986 [orig. 1925]. *The Theory of Peasant Economy*. Madison, Wisconsin: University of Wisconsin Press.

Chibnik, M. 1987. The economic effects of household demography: A cross-cultural assessment of Chayanov's theory. In M. D. Maclachlan, ed., *Household Economies and Their Transformations* pp. 74-106. Lanham, Maryland: University Press of America.

Chicchón, A. 1992. Chimane Resource Use and Market Involvement in the Beni Biosphere Reserve, Bolivia. Gainesville, Florida: Ph.D. diss., University of Florida.

CIDCA (Centro de Información y Documentación de la Costa Atlántica). 1982. *Demografía costeña. Notas sobre la historia demográfica y población actual de los grupos étnicos de la costa Atlántica Nicaragüense*. Managua: Centro de Información y Documentación de la Costa Atlántica.

Cleary, D. 1990. *Anatomy of the Amazon Gold Rush*. Iowa City, Iowa: University of Iowa Press.

Cleaver, K. M. and G. A. Schreiber. 1991. The population, agriculture and environment nexus in Sub-Saharan Africa. Washington, D.C.: The World Bank. Manuscript.

Coate, S. and M. Ravallion. 1993. Reciprocity without commitment: Characterization and performance of informal risk-sharing arrangements. *Journal of Development Economics* 40:1:1-24.

Collins, J. 1986. Smallholder settlement of tropical South America: The social causes of ecological destruction. *Human Organization* 45:1:1-10.

Confalonieri, U., L. F. Ferreira, and A. Araújo. 1991. Intestinal helminths in lowland South American Indians: Some evolutionary interpretations. *Human Biology* 63:6:863-873.

Conklin, B. A. and L. R. Graham. 1995. The shifting middle ground: Amazonian Indians and eco-politics. *American Anthropologist* 97:695-710.

Conzemius, E. 1932. *Ethnographical Survey of the Miskito and Sumu Indians of Honduras and Nicaragua*. Smithsonian Institution Bureau of American Ethnology, Bulletin 106. Washington, D.C.: Government Printing Office.

Cooper, R., C. Rotimi, S. Ataman, D. McGee, B. Osotimehin, S. Kadiri, W. Muna, S. Kingue, H. Fraser, T. Forrester, F. Bennett, and R. Wilks. 1997. The prevalence of hypertension in seven populations of West African origin. *American Journal of Public Health* 87:2:160-168.

Costa, D. L. 1997. Less of a luxury: The rise of recreation since 1888. Cambridge, Mass.: National Bureau of Economic Research Working Paper Series 6054.

——. 1998. The unequal work day: A long-term view. Cambridge, Mass.: National Bureau of Economic Research Working Paper Series 6419.

Cowgill, G. L. 1977. The trouble with significance tests and what we can do about it. *American Antiquity* 42:3:350-368.

Cropper, M. and C. Griffiths. 1994. The interaction of population growth and environmental quality. *American Economic Review* 84:2:250-254.

Cropper, M., S. K. Aydede, and P. R. Portney. 1992. Rates of time preference for saving lives. *American Economic Review* 82:2:469-473.

Cruz, G. and E. Benítez. 1994. *Diagnóstico etnológico y ecológico de la biósfera Tawahka Asangni*. Krausirpe, Biósfera Tawahka Asangni, Gracias a Dios, Honduras: Federación Indígena Tawahka de Honduras. 2 volumes.

Cuesta, M., G. Carlson, and E. Lutz. 1997. An empirical assessment of farmer's discount rates in Costa Rica. Washington, D.C.: The World Bank. Manuscript.

Daillant, I. 1994. Sens dessus-dissous. Organisation sociale et spatial des Chimanes d'Amazonie boliviene. Ph.D. diss., Laboratoire d'ethnologie et de Sociologie Comparative, Université de Paris.

Davidson, W. V. and F. Cruz. 1988. Delimitación de la región habitada por los sumos taguacas de Honduras 1600-1900. *Yaxkin* 1:123-136.

Davis, S. 1977. *Victims of the Miracle*. Cambridge: Cambridge University Press.

Deacon, R. T. 1994. Deforestation and the rule of law in a cross section of countries. *Land Economics* 70:4:414-417.

——. 1999. Deforestation and ownership: Evidence from historical accounts and contemporary data. *Land Economics*. 75:341-359.

Deaton, A. 1997. *The Analysis of Household Surveys. A Microeconomic Approach to Development Policy*. Baltimore, Maryland: The Johns Hopkins University Press.

DeFranco, M. and R. Godoy. 1992. The economic consequences of the coca industry in Bolivia: Historical, local, and macroeconomic perspectives. *Journal of Latin American Studies* 24:375-406.

——. 1993. Potato-led growth: The macroeconomic effects of technological innovations in Bolivian agriculture. *Journal of Development Studies* 29:3:561-587.

De Janvry, A. 1981. *The Agrarian Question and Reformism in Latin America*. Baltimore, Maryland: The Johns Hopkins University Press.

Demmer, J., R. Godoy, D. Wilkie, H. Overman, M. Taimur, K. Fernando, R. Gupta, T. Price, S. Sriram, N. Brokaw. 2001. Does economic development reduce the availability of wildlife? Waltham, Mass.: Sid-Heller, Braindeis University. Manuscript.

Denevan, W. H. 1973. Development and the imminent demise of the Amazon rain forest. *The Professional Geographer* 25:130-135.

——. 1992. The pristine myth: The landscape of the Americas in 1492. *Annals of the Association of American Geographers* 82:3:369-385.

DeWalt, B. R. and P. J. Pelto, eds. 1985. *Micro and Macro Levels of Analysis in Anthropology: Issues in Theory and Research*. Boulder, Colo.: Westview Press.

Diamond, J. 1994. Stinking birds and burning books. *Natural History*. 102:2:4-12.

Durrenberger, E. P. 1979. An analysis of Shan household production decisions. *Journal of Anthropological Research* 35:4:447-58.

Dwyer, P. D. and M. Minnegal. 1993. Are Kubo hunters 'show offs'? *Ethology and Sociobiology* 14:53-70.

Eggan, F. 1954. Social anthropology and the method of controlled comparison. *American Anthropologist* 56:743-763.

Ehui, S., T. W. Hertel, and P. V. Preckel. 1990. Forest resource depletion, soil dynamics, and agricultural productivity in the tropics. *Journal of Environmental Economics and Management* 18:136-154.

Ellis, F. 1988. *Peasant Economics: Farm Households and Agrarian Development.* Cambridge: Cambridge University Press.

Ellis, R. 1996. A Taste for Movement: An Exploration of the Social Ethics of the Tsimane´s of Lowland Bolivia. Ph.D. diss., St. Andrews University, Scotland.

———. 1999. Enseñemos a los tsimane´ a sumar monos: la inclusión de los indígenas (y exclusión de sus cosmologias) en el manejo de fauna. *Anales de la Reunión Anual de Etnología,* Museo de Etnología, La Paz. [in press].

Emmons, L. H. 1990. *Neotropical Rainforest Mammals. A Field Guide.* Chicago, Illinois: The University of Chicago Press.

EPRM (Equipo Pastoral Rural de Moxos). 1989. *Historia del pueblo de Moxos.* Volume 2. La Paz: Gráficos de Editorial Popular.

Fafchamps, M. 1992. Solidarity networks in preindustrial societies: Rational peasants with a moral economy. *Economic Development and Cultural Change* 41:1:147-174.

Falaris, E. and H. E. Peters. 1998. Survey attrition and schooling choices. *The Journal of Human Resources* 33:2:345-385.

Falconer, J. and J. E. M. Arnold. 1989. *Household Food Security and Forestry: An Analysis of Socioeconomic Issues.* Rome: FAO.

Falconer, J. and C. R. S. Koppell. 1990. *The Major Significance of Minor Forest Products. The Local Use and Value of Forests in the West African Humid Forest Zones.* Rome: FAO.

Ferguson, R. B. 1992. A savage encounter: Western contact and the Yanomami war complex. In R. B. Ferguson and N. Whitehead, eds., *War in the Tribal Zone: Expanding States and Indigenous Warfare,* pp. 199-227. Santa Fe, New Mexico: School of American Research Press.

———. 1995. *Yanomami Warfare: A Political History.* Santa Fe, New Mexico: School of American Research Press.

Finegan, B. 1996. Pattern and process in neotropical secondary rain forests: The first 100 years of succession. *Trends in Ecology and Evolution (TREE)* 11:3:119-124.

Fitzgerald, J., P. Gottschalk, and R. Moffitt. 1998. An analysis of sample attrition in panel data. *The Journal of Human Resources* 33:2:320-344.

Fleming-Morán, M., R. V. Santos, and C. E. A. Coimbra. 1991. Blood pressure levels of the Suruí Indians of the Brazilian Amazon: Group- and sex-specific effects resulting from body composition, health status, and age. *Human Biology* 63:6:835-861.

Foster, A. D. 1995. Prices, credit markets, and child growth in low-income rural areas. *Economic Journal* 105:430:551-571.

Foster, A. D. and M. R. Rosenzweig. 1995. Learning by doing and learning from others. Human capital and technical change in agriculture. *Journal of Political Economy* 103:6:1176-1210.

Foster, A. D., M. R. Rosenzweig, and J. R. Behrman. 1998. Population Growth, Income Growth and Deforestation: Management of Village Common Land in India. Providence, Rhode Island: Department of Economics, Brown University. Manuscript.

Foweraker, J. 1981. *The Struggle for Land: A Political Economy of the Pioneering Frontier in Brazil from 1930 to the Present Day.* New York: Cambridge University Press.

Frumhoff, P. C. 1995. Conserving wildlife in tropical forests managed for timber. *BioScience* 45:7:456-464.

Garro, L. C. 1986. Intracultural variation in folk medical knowledge: A comparison between curers and non-curers. *American Anthropologists* 88:2:351-370.

Gersovitz, M. 1988. Savings and development. In H. Chenery and T. N. Srinivasan, eds., *Handbook of Development Economics*, pp. 382-424. Amsterdam: Elsevier Science Publishers B.V.

Ghez, G. R. and G. S. Becker. 1975. *The Allocation of Time and Goods Over the Life Cycle.* New York: Columbia University Press.

Gibson, B. and R. Godoy. 1993. The distributive effects of Bolivia's drug dependence: A general-equilibrium approach. *World Development* 21:6:1007-1021.

Glewwe, P., M. Kremer, and S. Moulin. 1998. Textbooks and test scores: Evidence from a prospective evaluation in Kenya. Washington, D.C.: World Bank. Manuscript.

Godoy, R. 1994. The effect of rural education on the use of the forest by the Sumu: Pathways, qualitative findings, and policy guidelines. *Human Organization* 53:3:233-244.

——. 1999. The difference a year makes. *Science* 285:1850.

Godoy, R. and R. Lubowski. 1992. Guidelines for the economic valuation of non-timber tropical forest products. *Current Anthropology* 33:4:423-433.

Godoy, R., N. Brokaw, and D. Wilkie. 1995. The effect of income on the extraction of non-timber tropical forest products: Model, hypotheses, and preliminary findings from the Sumu Indians of Nicaragua. *Human Ecology* 23:1:29-52.

Godoy, R., N. Brokaw, D. Wilkie, G. Cruz, A. Cubas, J. Demmer, K. McSweeney, and J. Overman. 1996. Rates of return on investment in cattle among Amerindians of the rain forest of Honduras. *Human Ecology* 24:3:395-399.

Godoy, R., D. Wilkie, and J. Franks. 1997. The effects of markets on neotropical deforestation: A comparative study. *Current Anthropology* 38:5:875-8.

Godoy, R., D. Wilkie, K. O'Neill, P. Kostishack, S. Groff, J. Overman, J. Demmer, A. Cubas, K. McSweeney, and M. Martínez. 1997. Household determinants of neotropical deforestation by Amerindians in Honduras. *World Development* 25: 6: 977-987.

Godoy, R., M. Jacobson, J. DeCastro, V. Aliaga, J. Romero, and A. Davis. 1998. The role of tenure and private time preference in neotropical deforestation. *Land Economics* 74:2:162-170.

Godoy, R., M. Jacobson, and D. Wilkie. 1998. Strategies of rain-forest dwellers against misfortunes: The Tsimane´ Indians of Bolivia. *Ethnology* 37:1:55-69.

Godoy, R., P. Kostishack, D. Wilkie, and K. O'Neill. 1998a. The socioeconomic correlates of error in estimation of agricultural field size: An experimental study among the Tawahka Indians of Honduras. *Cultural Anthropology Methods Journal* 10:3:48-53.

Godoy, R., J. R. Franks, and M. Alvarado. 1998. Adoption of modern agricultural technologies by lowland indigenous groups in Bolivia: The role of households, villages, ethnicity, and markets. *Human Ecology* 26:3:351-369.

Godoy, R. and K. Kirby. 2000. A test of the Becker-Mulligan hypothesis of endogenous time preference: Evidence from four primitive societies in Bolivia. Waltham, Mass.: Department of Anthropology, Brandeis University. Manuscript.

Godoy, R., K. Kirby, and D. Wilkie. 2001. Tenure security, private time preference, and the use of natural resources among lowland Bolivian Amerindians. *Ecological Economics* [in press].

Godoy, R., K. Kirby, and D. Wilkie. 2001. Can consumer demand for bushmeat be reduced? A preliminary study of income and price elasticities of demand in lowland Amerindian societies. *Conservation Biology* [in press].

Godoy, R. and M. Contreras. 2001. A comparative study of schooling and tropical deforestation in Amerindian societies: Forest values, environmental externalities, and policy options. *Economic Development and Cultural Change* [in press].

Godoy, R. and. G. Wong. 1999. Does economic development increase vulnerability? A test of consumption smoothing among the Tawahka Indians of the Honduran rain forest. Waltham, Mass.: department of anthropology, Brandeis University. Manuscript.

Goland, C. 1993. Field scattering as agricultural risk management: A case study from Cuyo Cuyo, department of Puno, Peru. *Mountain Research and Development* 13:4:317-338.

Goleman, D. 1991. Shamans and their lore may vanish with forests. *New York Times*. June 11, 1991.

Gómez-Pompa, A., C. Vazques-Yáñez, and S. Guevara. 1972. The tropical rain forest: A non-renewable resource. *Science* 177:762-65.

Gray-Molina, G., W. Jiménez, E. Pérez de Rada, and E. Yáñez. 1998. Activos y recursos de la población pobre en latinoamerica: El caso de Bolivia. Washington, D.C.: Inter American Development Bank. Manuscript.

Green, L., A. Fry, and J. Myerson. 1994. Discounting of delayed rewards: A life-span comparison. *Psychological Science* 5:1:33-36.

Greenbaum, L. 1989. Plundering the timber on Brazilian Indian reservations. *Cultural Survival Quarterly* 13:1:23-26.

Gregory, C. A. 1982. *Gifts and Commodities*. London: Academic Press.

Griffin, K. 1979. *The Political Economy of Agrarian Change. An Essay on The Green Revolution*. London: MacMillan Press.

Griliches, Z. 1986. Economic data issues. In Z. Griliches and M. D. Intriligator, eds., *Handbook of Econometrics*, vol. 23, chapter 25. New York: North-Holand Publishing Company.

Gross, D. R., G. Eiten, N. M. Flowers, F. Leoi, M. L. Ritter, and D. W. Werner. 1979. Ecology and acculturation among native peoples of Central Brazil. *Science* 206:30:1043-1050.

Grossman, G. and A. Krueger. 1995. Economic growth and the environment. *Quarterly Journal of Economics* 34:353-77.

Gullison, R. E. 1995. Conservation of Tropical Forests Through the Sustainable Production of Forest Products: The Case of Mahogany (*Swietenia macrophylla*) in the Chimanes Forest, Beni, Bolivia. Ph.D. diss., Princeton University.

Gullison, R. E., S. N. Panfil, J. J. Strouse, and S. P. Hubbell. 1996. Ecology and management of mahogany (*Swietenia macrophylla* King) in the Chimanes Forest, Beni, Bolivia. *Botanical Journal of the Linnean Society* 122:1:9-34.

Haggblade, S., P. Hazell, and J. Brown. 1989. Farm-nonfarm linkages in rural Sub-Saharan Africa. *World Development* 17:1173-1201.

Hames, R. 1987. Garden labor exchange among the Ye'kwana. *Ethology and Sociobiology* 8:4:259-284.

——. 1990. Sharing among the Yanomamö: Part I, the effects of risk. In E. Cashdan, ed., *Risk and Uncertainty in Tribal and Peasant Economies*, pp. 89-106. Boulder, Colo.: Westview Press.

Harris, M. 1997. Criticism and interpretation. *Current Anthropology* 38:3:410-415.

Hausman, J. A. 1979. Individual discount rates and the purchase and utilization of energy-using durables. *Bell Journal of Economics* 10:1:33-54.

Hawkes, K. 1992a. Sharing and collective action. In E. Smith and B. Winterhalder, eds., *Evolutionary Ecology and Human Behavior*, pp. 269-300. New York: Aldine.

——. 1992b. On sharing and work. *Current Anthropology* 33:4:404-407.

Hawkes, K., H. Kaplan, K. Hill, and A. M. Hurtado. 1987. Aché at the settlement: Contrasts between farming and foraging. *Human Ecology* 15:2:133-161.

Hazell, P. B. R. and C. Ramsamy, eds. 1991. *The Green Revolution Reconsidered. The Impact of High-Yielding Rice Varieties in South India.* Baltimore, Maryland: The Johns Hopkins University Press.

Hecht, S. B. and A. Cockburn. 1989. *The Fate of the Forest: Developers, Destroyers, and Defenders of the Amazon.* New York: Harper Perennia.

Hecht, S. B. 1985. Environment, development and politics: Capital accumulation and the livestock sector in eastern Amazonia. *World Development* 13:6:663-684.

——. 1993. The logic of livestock and deforestation in Amazonia. *BioScience* 43:10:687-695.

——. 1998. When solutions become drivers: The dynamics of deforestation in Bolivia. Los Angeles, California: University of California, Los Angeles, Department of Urban Studies. Manuscript.

Helms, M. W. 1968. The cultural ecology of a colonial tribe. *Ethnology* 8:76-84.

——. 1968a. Matrilocality and the maintenance of ethnic identity: The Miskito of eastern Nicaragua and Honduras. *37th International Congress of Americanists*. Stuttgart. Volume 2, pp. 459-464.

——. 1971. *Asang. Adaptations to Cultural Contact in a Miskito Community.* Gainesville, Florida: University of Florida Press.

Herlihy, P. H. and A. P. Leake. 1990. The Tawahka Sumu: A delicate balance in Mosquitia. *Cultural Survival Quarterly* 14:4:13-16.

——. 1991. Propuesta reserva forestal Tawahka Sumu. Tegucigalpa: Instituto Hondureño de Antropología e Historia and Mosquitia Pawisa (MOPAWI). Manuscript.

——. 1992. Situación actual del frente de colonización/deforestación en la región propuesta para el parque nacional Patuca. Tegucigalpa: Mosquitia Pawisa (MOPAWI). Manuscript.

——. 1997. Indigenous peoples and biosphere reserve conservation in the Mosquitia rain forest corridor, Honduras. In Stanley F. Stevens, ed., *Conservation Through Cultural Survival: Indigenous Peoples and Protected Areas.* Washington, D.C.: Island Press, pp. 99-129.

Hill, K. and H. Kaplan. 1993. On why male foragers hunt and share food. *Current Anthropology* 34:5:701-706.

Hill, K. and K. Hawkes. 1983. Neotropical hunting among the Aché of eastern Paraguay. In R. B. Hames and W. T. Vickers, eds., *Adaptive Responses of Native Amazonians*, pp.139-188. New York: Academic Press.

Hill, K. and M. Hurtado. 1996. *Aché Life History. The Ecology and Demography of a Foraging People.* New York: Aldine De Gruyter.

Hill, M. S. 1985a. Investments of time in houses and durables. In F. T. Juster and F. P. Stafford, eds., *Time, Goods, and Well-Being*, pp. 177-204. Ann Arbor, Michigan: Institute for Social Research, University of Michigan.

——. 1985b. Patterns of time use. In F. T. Juster and F. P. Stafford, eds., *Time, Goods, and Well-Being*, pp. 133-176. Ann Arbor, Michigan: Institute for Social Research, University of Michigan.

Holden, S. T., S. Berkele, and M. Wilk. 1998. Poverty, market imperfection, and time preference: Of relevance for environmental policy? *Environment and Development Economics* 3:1:105-30.

Holmes, R. 1985. Nutritional status and cultural change in Venezuela's Amazon territory. In J. Hemming, ed., *Change in the Amazon Basin. Volume II. The Frontier After a Decade of Colonization*, pp. 237-255. Manchester: Manchester University Press.

Homma, A. K. O. 1992. The dynamics of extraction in Amazonia: A historical perspective. *Advances in Economic Botany* 9:42-65.

House, P. R. 1997. *Farmers of the Forest*. London: The Natural History Museum.

Howard, A. F., R. E. Rice, and R. E. Gullison. 1996. Simulated financial returns and selected environmental impacts from four alternative silvicultural prescriptions applied in the neotropics: A case study of the Chimanes Forest, Bolivia. *Forest Ecology and Management* 89:1-3:43-57.

Huanca, T. 1999. Tsimane´ Indigenous Knowledge, Swidden Fallow Management, and Conservation. Ph.D. diss., University of Florida, Gainesville.

Huber, J. R. 1971. Effect on prices of Japan's entry into world commerce after 1858. *Journal of Political Economy* 79:3:614-628.

Hurrell, A. 1990. The politics of Amazonian deforestation. *Journal of Latin American Studies* 23:197-215.

Hyde, W. F., G. S. Amacher, and W. Magrath. 1995. Deforestation and forest land use: Theory, empirical evidence, and policy implications. *The World Bank Research Observer* 11:2:223-248.

Institute of Medicine, 1997: *America's Vital Interest in Global Health: Protecting our People, Enhancing our Economy, and Advancing our International Interests*. Washington, DC: National Academy Press.

Irvine, D. 1989. Succession management and resource distribution in an Amazonian rain forest. *Advances in Economic Botany* 7:223-237.

Jenkins, C. 1989. Culture change and epidemiological patterns among the Hagahai, Papua New Guinea. *Human Ecology* 17:1:27-57.

Jenkins, J. 1986. *El desafío indígena en Nicaragua: El caso de los Miskitos*. Mexico: Editorial Katún.

Jirström, M. 1996. *In the Wake of the Green Revolution: Environmental and Socio-Economic Consequences of Intensive Rice Agriculture - The Problems of Weeds in Muda, Malaysia*. Lund: Lund University Press.

Johansson, S. R. 1991. The health transition: The cultural inflation of morbidity during the decline of mortality. *Health Transition Review* 1:1:39-68.

Johnston, B. F. and P. Kilby. 1975. *Agriculture and Structural Transformation.* London: Oxford University Press.

Jones, J. 1980. Conflict Between Whites and Indians in the Llanos de Moxos, Beni Department: A Case Study in Development from the Cattle Regions of the Bolivian Oriente. Ph.D. diss., University of Florida.

——. 1991. *Economic, Political Power, and Ethnic Conflict on a Changing Frontier: Notes from the Beni Department in Eastern Bolivia.* Binghamton, New York: Institute for Development Anthropology, Working Paper #58.

——. 1995. Environmental destruction, ethnic discrimination, and international aid in Bolivia. In M. Painter and W. H. Durham, eds., *The Social Causes of Environmental Destruction in Latin America*, pp. 169-216. Ann Arbor, Michigan: University of Michigan Press.

Jorgenson, J. P. 1997. Cambios en los patrones de la caceria de subsistencia a través de mejoramientos socio-económicos: el ejemplo de los cazadores Maya en México. In T. G. Fang, R. E. Bodmer, R. Aquino, and M. H. Valqui, eds., *Manejo de fauna silvestre en la amazonía*, pp. 31-40. La Paz: OFAVIM.

Juster, F. T. and F. P. Stafford. 1991. The allocation of time: Empirical findings, behavioral models, and problems of measurement. *Journal of Economic Literature* 29:471-522.

Kaja, F. 1984. Non-sharing of medical knowledge among specialist healers and their patients: A contribution to the study of intracultural diversity and practitioner-patient relations. *Medical Anthropology* 8:3:195-209.

Kaplan, H. and K. Hill. 1985. Food sharing among Aché foragers: Tests of explanatory hypotheses. *Current Anthropology* 26:2:223-246.

Kennedy, P. 1993. *A Guide to Econometrics.* Cambridge, Mass.: MIT Press.

Kennedy, E. 1994. Health and nutrition effects of commercialization of agriculture. In J. von Braun and E. Kennedy, eds., *Agricultural Commercialization, Economic Development, and Nutrition*, pp. 79-99. Baltimore, Maryland: The Johns Hopkins University Press.

King, G. 1997. *A Solution to the Ecological Inference Problem. Reconstructing Individual Behavior from Aggregate Data.* Princeton, New Jersey: Princeton University Press.

Kirby, K. N. and N. N. Marakovic. 1996. Delay-discounting probabilistic rewards: Rates decrease as amounts increase. *Psychonomic Bulletin & Review* 3:1:100-104.

Kranton, R. E. 1996. Reciprocal exchange: A self-sustaining system. *American Economic Review* 86:4:830-851.

Krekeler, B. 1995. *Historia de los chiquitanos*. La Paz: Talleres Gráficos HISBOL.

Kroeger, A. and F. Barbira-Freedman. 1982. *Cultural Change and Health: The Case of South American Rain Forest Indians*. Frankfurt: Verlag Peter Land. Volume 12.

Krugman, A. and S. Obstfeld. 1997. *International Economics: Theory and Policy*. New York: Addison-Wesley.

Lambert, S. M. 1931. Health survey of Rennell and Bellona islands. *Oceania* 2:2:136-173.

Landero, F. 1935. Los Taoajkas o Sumos del Patuca y Wampú. *Anthropos* 30:33-50.

Lawrance, E. C. 1991. Poverty and the rate of time preference: Evidence from panel data. *Journal of Political Economy* 99:54-77.

Lawrence, D., M. Leighton, and D. Peart. 1995. Availability and extraction of forest products in primary and managed forests around a Dyak village in West Kalimantan, Indonesia. *Conservation Biology* 9:76-81.

Leatherman, T. L. 1994. Health implications of changing agrarian economies in the southern Andes. *Human Organization* 53:4:371-380.

Leatherman, T. L., J. W. Carey, and R. B. Thomas. 1995. Socioeconomic change and patterns of growth in the Andes. *American Journal of Physical Anthropology* 97:307-321.

Lehm, Z. 1991. Loma Santa: procesos de reducción, dispersión y reocupación del espacio de los indígenas Mojeños. Manuscript.

——. 1994. Estudio socio económico del territorio indígena Reserva de la Biósfera Pilón Lajas. La Paz: Veterinarios Sin Fronteras and CIDDEBENI.

Leonard, W. R., K. M. Dewalt, J. E. Uquillas, and B. R. Dewalt. 1994a. Ecological correlates of dietary consumption and nutritional status in highland and coastal Ecuador. *Ecology of Food and Nutrition* 31:67-85.

——. 1994b. Diet and nutritional status among cassava producing agriculturalists of coastal Ecuador. *Ecology of Food and Nutrition* 32:113-127.

Levins, R. 1994. The emergence of new diseases. *American Scientists* 85:52-60.

Lévy-Strauss, C. 1969. *The Elementary Structures of Kinship*. Boston, Mass.: Beacon Press.

Lindert, P. H. 1991. Historical patterns of agricultural policy. In C. P. Timmer, ed., *Agriculture and the State: Growth, Employment, and Poverty in Developing Countries*, pp. 29-83. Ithaca: Cornell University Press.

Lipton, M. and R. Longhurst. 1989. *New Seeds and Poor People*. London: Unwin Hyman.

Locay, L. 1990. Economic development and the division of production between households and markets. *Journal of Political Economy* 98:5:1:965-982.

Loewenstein, G. 1992. The fall and rise of psychological explanations in the economics of intertemporal choice. In G. Loewenstein and J. Elster, eds., *Choice Over Time*, pp. 3-34. New York: Russell Sage Foundation.

Lommitz, L. A. 1977. *Networks and Marginality: Life in a Mexican Shantytown*. New York: Academic Press.

López, R. 1986. Structural models of the farm household that allow for interdependent utility and profit maximization decisions. In I. Sing, L. Squire, and J. Strauss, eds., *Agricultural Household Models: Extensions, Applications and Policy*, pp. 306-326. Baltimore, Maryland: Johns Hopkins University Press.

——. 1993a. Resource degradation, community controls and agricultural production in tropical areas. College Park, Maryland: Department of Economics, University of Maryland. Manuscript.

——. 1993b. Economy-wide policies, agricultural productivity, and environmental factors: The case of Ghana. College Park, Maryland: Department of Economics, University of Maryland. Manuscript.

Lugo, A. 1992. Tropical forest use. In T. E. Downing, S. B. Hecht, H. A. Pearson, and C. Garcia-Downing, eds., *Development or Destruction? The Conversion of Tropical Forest to Pasture in Latin America*, pp. 117-132. Boulder, Colo.: Westview Press.

Mace, R. 1990. Pastoralist herd composition in unpredictable environments: A comparison of model predictions and data from camel-keeping groups. *Agricultural Systems* 33:1-11.

Mace, R. and A. I. Houston. 1989. Pastoralist strategies for survival in unpredictable environments: A model of herd composition that maximizes household viability. *Agricultural Systems* 31:185-204.

MacLeod, M. 1979. Forms and types of work, and the acculturation of the colonial Indian of Mesoamerica: Some preliminary observations. In E. C. Frost, M. C. Meyer, and J. Z. Vásquez, eds., *Labor and Laborers Through Mexican History*, pp. 75-92. Tucson, Arizona: University of Arizona Press.

Malinowski, B. 1961 [orig. 1922]. *Argonauts of the Western Pacific.* New York: Dutton.

Malkin, B. 1956. Sumu ethnozoology: Herpetological knowledge. *Davidson Journal of Anthropology* 2:2:165-180.

Marshall, A. 1936. *Principles of Economics.* London: MacMillan.

Marshall, L. 1961. Sharing, talking, and giving: Relief of social tensions among the Kung Bushmen. *Africa* 29:335-65.

Mather, K. F. 1922. Exploration in the land of the Yuracarés, eastern Bolivia. *Geographical Review* 12:42-56.

Mauss, M. 1990 [orig. 1927]. *The Gift: The Form and Reason For Exchange in Archaic Societies.* New York: W. W. Norton.

McCloskey, D. and S. Ziliak. 1996. The standard error of regressions. *Journal of Economic Literature* 34:97-114.

McDaniel, J. 2000. Indigenous Organizations and Conservation/Development Organization: The Politics of Ethnicity. Ph.D. diss. University of Florida, Gainesville.

McSweeney, K. 1999. The canoe in the tree. *Equinox.* 103:78-86.

Mellor, J. W. 1976. *The New Economics of Growth.* Ithaca: Cornell University Press.

——. 1988. Food policy, food aid, and structural adjustment programs: The context of agricultural development. *Food Policy* 13:1:10-17.

Métraux, A. 1948. The Yuracaré, Mosetene, and Chimane. In J. H. Steward, ed., *Handbook of South American Indians*, Vol. 3, pp. 23-48. Washington: Smithsonian Institution, Bureau of American Ethnology, Bulletin 143.

Miller, L. E. 1917. The Yuracaré Indians of eastern Bolivia. *Geographical Review* 4:6:450-464.

Minge-Kalman, W. 1977. On the theory and measurement of domestic labor intensity. *American Ethnologist* 4: 273-84.

Minnegal, M. 1996. A necessary unity: The articulation of ecological and social explanations of behavior. *The Journal of the Royal Anthropological Institute* 2:1:141-158.

Mischel, W., Y. Shoda, and M. L. Rodríguez. 1989. Delay of gratification in children. *Science* 244:4907:933-938.

Molina, W. 1994. *TIPNIS hoy.* Trinidad: CIDDEBENI, publicación #28.

Morán, E. 1993a. *Through Amazonian Eyes: The Human Ecology of Amazonian Populations.* Iowa City, Iowa: University of Iowa Press.

——. 1993b. Deforestation and Land Use in the Brazilian Amazon. *Human Ecology* 21:1:1-21.

Morduch, J. 1995. Income smoothing and consumption smoothing. *Journal of Economic Perspectives* 9:3:103-114.

——. 1998. Between the market and state: Can informal insurance patch the safety net? Princeton, New Jersey: Wooddrow Wilson, Princeton University. Manuscript.

Moser, C. 1998. The asset-vulnerability framework: Reassessing urban poverty reduction strategies. *World Development* 26:1:1-19.

Murphy, R. F. 1956. Credit versus cash. *Human Organization* 14:3:26-30.

——. 1960. *Headhunter's Heritage: Social and Economic Change Among the Mundurucú Indians.* Berkeley, California: University of California Press.

——. 1961-1962. Deviancy and social control. *Kroeber Anthropological Society Papers* 24:55-61, 27:49-54.

Murphy, R. F. and J. R. Steward. 1956. Tappers and trappers: Parallel processes in acculturation. *Economic Development and Cultural Change* 4:335-355.

Murray, C. J. L. and L. C. Chen. 1992. Understanding morbidity change. *Population and Development Review* 18:3:481-503.

Murray, C. J. L., C. Yang, and X. Qiao. 1992. Adult mortality: Levels, patterns, and causes. In R. G. A. Freachem, T. Kjellstrom, C. J. L. Murray, M. Over, and M. A. Phillips, eds., *The Health of Adults in the Developing World*, pp. 23-112. Oxford: Oxford University Press.

Netting, R. M. 1982. Some home truths on household size and wealth. *American Behavioral Scientist* 25:6:641-62.

——. 1993. *Smallholders, Householders: Farm Families and the Ecology of Intensive, Sustainable Agriculture.* Stanford, California: Stanford University Press.

Nickson, A. 1995. *Local Government in Latin America.* London: Lynne Rienner.

Norgan, N. G. 1994. Interpretation of low body-mass indices: Australian aborigines. *American Journal of Physical Anthropology* 94:229-237.

Nordenskiold, E. 1924. *The Ethnography of South America Seen From Mojos in Bolivia.* Göteborg: Comparative Ethnographical Studies No. 3.

Ogle, B. 1996. People's dependency on forests for food security. Some lessons learned from a program of case studies. In M. Ruiz Pérez and J. E. M. Arnold, eds., *Current Issues in Non-Timber Forest Products Research*, pp. 219-241. Bogor, Indonesia: Center for International Forestry Research (CIFOR).

O'Meara, T. 1997. Causation and the struggle for a science of culture. *Current Anthropology* 38:3:399-410.

Packard, R. M. and P. J. Brown. 1997. Rethinking health, development, and malaria: Historicizing a cultural model in international health. *Medical Anthropology* 17:181-194.

Padilla-Lobo, A. 1995. Estrategias de subsistencia de los indígenas Tawahkas en la comunidad de Krausirpe, Honduras. B. A. thesis, Universidad Austral de Chile, Valdivia, Chile.

Painter, M. and W. H. Durham, eds. 1995. *The Social Causes of Environmental Destruction in Latin America*. Ann Arbor, Michigan: University of Michigan Press.

Parry, J. 1993. The gift, the Indian gift, and the 'indian gift'. *Man* 21:3:453-473.

Parry, J. and M. Bloch, eds. 1989. *Money and the Morality of Exchange*. Cambridge: Cambridge University Press.

Patel, S. H., T. C. Pinckney, and W. K. Jaeger. 1995. Smallholder wood production and population pressure in East Africa: Evidence of an environmental Kuznets curve? *Land Economics* 71:4:516-30.

Paxson, C. 1992. Using weather variability to estimate the response of savings to transitory income in Thailand. *American Economic Review* 82:1:15-33.

Paz Patiño, S. 1991. Conflictos interétnicos en la región del habitat tradicional Yuracaré. Cochabamba: Carrera de sociología, facultad de ciencias económicas y sociales, Universidad Mayor de San Simón.

Pender, J. L. 1996. Discount rates and credit markets. Theory and evidence from rural India. *Journal of Development Economics* 50:2:257-297.

Pérez-Diez, A. 1984. *Urgent Anthropological Research Among the Chimane (Eastern Bolivia)*. Bulletin of the International Committee on Urgent Anthropological and Ethnological Research No. 26.

Picchi, D. 1991. The impact of an industrial agricultural project on the Bakairi Indians of Central Brazil. *Human Organization* 50:1:26-38.

Piland, R. A. 1991. Traditional Chimane Agriculture and Its Relation to Soils of the Beni Biosphere Reserve, Bolivia. M.A. thesis, University of Florida.

Pitt, M. and M. Rosenzweig. 1986. Agricultural prices, food consumption, and the health and productivity of Indonesian farmers. In I. Sing, L. Squire, and J. Strauss, eds., *Agricultural Household Models: Extensions, Applications and Policy*, pp. 153-182. Baltimore, Maryland: Johns Hopkins University Press.

Plotkin, M. J. 1993. *Tales of a Shaman's Apprentice: An Ethnobotanist Searches for New Medicines in the Amazon Rain Forest*. New York: Viking Press.

Posey, D. A. 1984. A preliminary report on diversified management of tropical forest by Kayapó Indians of the Brazilian Amazon. *Advances in Economic Botany* 4:112-127.

Press, S. J. and S. Wilson. 1978. Choosing between logistic regression and discriminant analysis. *Journal of the American Statistical Association* 73:699-705.

Redfield, R. 1941. *The Folk Cultures of Yucatan*. Chicago, Illinois: The University of Chicago Press.

——. 1947. The folk society. *American Journal of Sociology* 52:4:295-98.

Reed, R. 1995. Household ethnicity, household consumption: Commodities and the Guaraní. *Economic Development and Cultural Change* 44:1:129-145.

Ribera, R. J. 1983. Aima Suñe. Un estudio de la situación actual de la etnia Yuracaré del oriente Boliviano: Bibliografía e investigación de campo. Cochabamba: Facultad de Filosofía y Ciencias Religiosas, Universidad Católica Boliviana.

Ribot, J. C. 1993. Forestry policy and charcoal production in Senegal. *Energy Policy* 21:5:559-586.

Rice, R. E., R. E. Gullison, and J. W. Reid. 1997. Can sustainable management save tropical forests? *Scientific American* 42(April): 44-49.

Rich, E. E. 1960. Trade habits and economic motivation among the Indians of North America. *Canadian Journal of Economics and Political Science* 26:1:35-53.

Ridgely, R. S. and J. A. Gwynne Jr. 1989. *Birds of Panama, with Costa Rica, Nicaragua, and Honduras*. Princeton, New Jersey: Princeton University Press.

Riester, J. 1976. *En busca de la loma santa*. La Paz: Los Amigos del Libro.

——. 1993. *Universo mítico de los Chimane*. La Paz: Talleres Gráficos HISBOL.

Riester, J. and B. Suaznabar. 1990. Diagnóstico del area III. Santa Cruz. Pueblo Chiquitano. Santa Cruz: Manuscript.

Rioja, G. 1992. The jatata project: The pilot experience of Chimane empowerment. In M. J. Plotkin and L. M. Famolare, eds., *Sustainable Harvest and Marketing of Rain Forest Products*, pp. 192-196. Washington, D.C.: Island Press.

Robinson, J. and E. Bennet, eds. 1999. *Hunting for Sustainability in the Tropical Forests*. New York: Columbia University Press.

Rogers, A. R. 1994. Evolution of time preference by natural selection. *American Economic Review* 84:3:460-481.

Roper, J. M. 1999. The Political Ecology of Indigenous Self-Development in Bolivia's Multiethnic Indigenous Territory. Ph.D. diss., University of Pittsburgh, Pennsylvania.

Rose, E. 1994. Consumption smoothing and excess female mortality in rural India. Saint Louis, Missouri: Department of Economics, University of Washington. Manuscript.

Rosenzweig, M. R. 1988. Risk, implicit contracts and the family in rural areas of low-income countries. *Economic Journal* 98:4:1148-1170.

Rosenzweig, M. R. and K. Wolpin. 1993. Credit market constraints, consumption smoothing, and the accumulation of durable production assets in low-income countries: Investments in bullocks in India. *Journal of Political Economy* 101:223-244.

Rosenzweig, M. R., J. R. Behrman, and P. Vashishtha. 1995. Location-specific technical change, human capital and local economic development: The Indian green revolution experience. In H. Siebert, ed., *Locational Competition and the World Economy*, pp. 24-47. Kiel: Kiel Institute of Economics.

Rozo, B. 1999. El pueblo chiquitano. Ficha resumen de un pueblo indígena. La Paz, Bolivia: Department of Anthropology, Universidad Mayor de San Andrés. Manuscript.

Rudel, T. K. 1989. Population, development, and tropical deforestation: A cross-national study. *Rural Sociology* 54:3:327-338.

——. 1998. Is there a forest transition? Deforestation, reforestation, and development. *Rural Sociology* 63:4:533-552.

Rudel, T. K. and B. Horowitz. 1993. *Tropical Deforestation: Small Farmers and Land Clearing in the Ecuadorian Amazon*. New York: Columbia University Press.

Ruitenbeek, H. J. 1988. Social cost-benefit analysis of the Korup Project, Cameroon. London: World Wide Fund for Nature Publication 3206\A14.1.

——. 1989. Economic analysis of issues and projects relating to the establishment of the proposed Cross River National Park (Oban Division) and support zone. London: World Wide Fund for Nature, Oban Feasibility Study. Manuscript.

Sackett, R. 1996. Time, Energy, and the Indolent Savage. A Quantitative Cross-Cultural Test of the Primitive Affluence Hypothesis. Ph.D. diss., University of California, Los Angeles.

Sahlins, M. 1968. Notes on the original affluent society. In R. Lee and I. DeVore, eds., *Man the Hunter*, pp. 85-89. New York: Aldine Publishing Company.

——. 1971. The intensity of domestic production in primitive societies: Social inflections of the Chayanov slope. In G. Dalton, ed., *Studies in Economic Anthropology*, pp 30-51. Washington, D.C.: American Anthropological Society.

——. 1972. *Stone Age Economics*. Chicago, Illinois: Aldine-Atherton.

Saldarriaga, J. G., D. C. West, M. L. Tharp, and C. Uhl. 1985. Long-term chronosequence of forest succession in the Upper Rio Negro of Colombia and Venezuela. *Journal of Ecology* 76:938-958.

Santos, R. V. and C. E. A. Coimbra. 1991. Socioeconomic transition and physical growth of Tupí-Mondê Amerindian children in the Aipuanã Park, Brazilian Amazon. *Human Biology* 63:6:795-819.

——. 1996. Social differentiation and body morphology in the Suruí of southwestern Amazonia. *Current Anthropology* 37:5:851-856.

Santos, R. V., N. M. Flowers, C. E. A. Coimbra Jr., and S. A. Gugelmin. 1997. Tapirs, tractors, and tapes: The changing economy and ecology of the Xavánte Indians of Central Brazil. *Human Ecology* 25:4:545-566.

Schemo, D. J. 1999. The last tribal battle. *The New York Times Magazine*. October 31.

Schiff, M. and A. Valdés. 1992. *The Political Economy of Agricultural Price Policy*. Volume 4. Baltimore, Maryland: The Johns Hopkins University Press.

Schmink, M. and C. Wood. 1987. The 'political ecology' of Amazonia. In P. D. Little, M. M. Horowitz, and A. E. Nyerges, eds., *Lands at Risk in the Third World: Local-Level Perspectives*, pp. 38-57. Boulder, Colo.: Westview Press.

——. 1992. *Contested Frontiers in Amazonia*. New York: Columbia University Press.

Schuh, E. G. and A. S. P. Brandão. 1992. The theory, empirical evidence, and debate on agriculture development issues in Latin America: A selective survey. In Lee R. Martin, ed., *A Survey of Agricultural Economics Literature, Vol. 4. Agriculture in Economic Development, 1940s to 1990s*, pp. 545-963. Minneapolis, Minnesota: University of Minnesota Press.

Schwarz, B. 1993. Tendencies del uso de la tierra en el departamento de Santa Cruz de la Sierra. Estudios de caso. Informe de consultoria preliminar para la Comisión Internacional del Medio Ambiente. Santa Cruz: manuscript.

Scobie, G. M. and R. Posada. 1978. The impact of technological change on income distribution: The case of rice in Colombia. *American Journal of Agricultural Economics* 23:162-175.

Scoones, I and J. Thompson. 1994. Knowledge, power, and agriculture: Towards a theoretical understanding. In I. Scoones, J. Thompson, and R. Chambers, eds., *Beyond Farmer First*, pp. 16-31. London: Intermediate Technology Publications.

Scoones, I., M. Melnyk, and J. N. Pretty. 1992. *The Hidden Harvest: Wild Foods and Agricultural Systems. A Literature Review and Annotated Bibliography*. London: IUC Press.

Shanin, T. 1986. Chayanov's message: Illuminations, miscomprehensions and the contemporary 'development theory'. In D. Thorner, B. Kerblay, and R. E. F. Smith, eds., *The Theory of Peasant Economy*, pp. 1-24. Madison, Wisconsin: University of Wisconsin Press.

Shephard, R. J. and A. Rode. 1996. *The Health Consequences of "Modernization". Evidence from Circumpolar Peoples*. Cambridge: Cambridge University Press.

Sierra, R. and J. Stallings. 1998. The dynamics and social organization of tropical deforestation in north-west Ecuador, 1983-1995. *Human Ecology* 26:1:135-161.

Sillitoe, P. 1998. The development of indigenous knowledge. *Current Anthropology* 39:2:223-252.

Silver, W. L., S. Brown, and A. E. Lugo. 1996. Effects of changes in biodiversity on ecosystem function in tropical forests. *Conservation Biology* 10:1:17-24.

Singh, H. 1999. *Reaffirming Ruralism. Managing Green Revolutions in Contemporary Punjab*. Manuscript.

Siskind, J. 1973. *To Hunt in the Morning*. New York: Oxford University Press.

Smith, A. 1884. *An Inquiry into the Nature and Causes of the Wealth of Nations*. London: T. Nelson and Sons, Paternoster Row.

Smith, R. C. 1996. Biodiversity won't feed our children. Biodiversity conservation and economic development in indigenous Amazonia. In K. H. Redford and J. A. Mansour, eds., *Traditional Peoples and Biodiversity Conservation in Large Tropical Landscapes*, pp. 197-218. Arlington, Virginia: America Verde Publications

Smith, R. C. and C. Tapuy. 1995. The economic challenge for indigenous Amazonians: Reciprocity, collective action, and common property in a market economy. Lima: Oxfam America. Manuscript.

Southgate, D. 1991. Tropical deforestation and agricultural development in Latin America. International Institute for Environment and Development Paper DP 91-01. London: London Environmental Economic Center.

Sowell, T. 1998. *Conquest and Cultures. An International History.* New York: Basic Books.

Sponsel, L. E., T. N. Headland, and R. C. Bailey, eds. 1996. *Tropical Deforestation. The Human Dimension.* New York: Columbia University Press.

Stearman, A. M. 1990. The effects of settler incursion on fish and game resources of the Yuquí. A native Amazonian society of eastern Bolivia. *Human Organization* 49:4:373-385.

Stern, D. I., M. S. Common, and E. E. Barbier. 1996. Economic growth and environmental degradation: The environmental Kuznets curve and sustainable development. *World Development* 24:7:1151-1160.

Stier, F. 1982. Domestic economy: Land, labor, and wealth in a San Blas community. *American Ethnologist* 9:3:519-537.

——. 1983. Modeling migration: Analyzing migration historically from a San Blas Cuna community. *Human Organization* 42:1:9-22.

Strauss, J. and D. Thomas. 1988. Nutrient intakes and income. In H. Chenery and A. Srinivasan, eds., *Handbook of Development Economics*, pp. 1893-1908. Amsterdam: North Holland.

Strauss, J., P. J. Gertler, O. Rahman, and K. Fox. 1993. Gender and life-cycle differentials in the patterns and determinants of adult health. *The Journal of Human Resources* 28:4:791-837.

Tannenbaum, N. 1984. The misuse of Chayanov: Chayanov's rule and empiricist bias in anthropology. *American Anthropologist* 86: 924-42.

Thaler, R. 1981. Some empirical evidence on dynamic inconsistency. *Economics Letters* 8:3:201-207.

Thaler, R. and G. Loewenstein. 1989. Intertemporal choice. *Journal of Economic Perspectives* 3:181-193.

Thiele, G., J. Johnson, and J. Wadsworth. 1995. *Bosquejo socioeconómico del norte de Bolivia.* Santa Cruz: Centro de Investigación Agrícola Tropical and Misión Británica en Agricultura Tropical. Informe Técnico No. 20.

Thirakul, S. nd. *Manual de dendrología del bosque latifoliado.* Tegucigalpa, Honduras: Corporación Hondureña de Desarrollo Forestal.

Time. 1991. Lost tribes, lost knowledge. September 23.

Timmer, P. C. 1995. Getting agriculture moving: Do markets provide the right signals? *Food Policy* 20:5:455-472.

Townsend, R. M. 1994. Risk and insurance in village India. *Econometrica* 62:3:539-591.

———. 1995. Consumption insurance: An evaluation of risk-bearing systems in low-income economies. *Journal of Economic Perspectives* 9:3:83-102.

Townson, I. M. 1994. Forest products and household incomes: A review and annotated bibliography. Oxford: Oxford University. Manuscript.

Tripp, R. 1996. Biodiversity and modern crop varieties: Sharpening the debate. *Agriculture and Human Values* 13:48-63.

Udry, C. 1994. Risk and insurance in a rural credit market: An empirical investigation in northern Nigeria. *Review of Economic Studies* 61:3:495-526.

———. 1995. Risk and saving in northern Nigeria. *American Economic Review* 85:5:1287-1301.

Vandana, S. 1992. *The Violence of the Green Revolution.* London: Zed Books Ltd.

Vickers, W. T. 1988. Game depletion hypothesis of Amazonian adaptation: Data from a native community. *Science* 239: 1521-1522.

———. 1994. From opportunism to nascent conservation. The case of the Siona-Secoya. *Human Nature* 5: 4: 307-337.

Vidal, L. 1989. Questao indígena e meio ambiente. Ambates entre culturas e interesses diferenciados. *Sao Paulo em Perspectiva* 3:4:50-55.

von Braun, J. and E. Kennedy, eds. 1994. *Agricultural Commercialization, Economic Development, and Nutrition.* Baltimore, Maryland: The Johns Hopkins University Press.

Wagley, C. 1955. Tapirapé Social and Cultural Change. 1940-1953. *31st International Congress of Americanists.* August 23-28, 1954. Volume 1. São Paulo: Editorial Anhem.

Walsh, J. F., D. H. Molyneux, and M. H. Birley. 1993. Deforestation: Effects on vector-borne disease. *Parasitology* 106:S55-S75.

Watts, M. 1987. Drought, environment, and food security: Some reflections on peasants, pastoralists and commoditization in dryland West Africa. In M. H. Glantz, ed., *Drought and Hunger in Africa: Denying Famine a Future,* pp. 171-212. New York: Cambridge University Press.

Werner, D. 1979. Subsistence productivity and hunting effort in native South America. *Human Ecology* 7:4:303-316.

Wilkie, D. S. and R. A. Godoy. 1996. Trade, indigenous rain forest economies and biological diversity. Model predictions and directions for research. In M. R. Pérez and J. E. M. Arnold, eds., *Current Issues in Non-Timber Forest Products Research*, pp. 83-102. Bogor, Indonesia: Center for International Forestry Research (CIFOR).

Wilmsen, E. N. 1989. *Land Filled with Flies: A Political Economy of the Kalahari.* Chicago, Illinois: The University of Chicago Press.

Wilson, E. O. and R. H. MacArthur. 1967. *The Theory of Island Biogeography.* Princeton, New Jersey: Princeton University Press.

Wilson, M. and M. Daly. 1997. Life expectancy, economic inequality, homicide, and reproductive timing in Chicago neighborhoods. *British Medical Journal* 314:1271-1274.

Wilson, M., M. Daly, and S. Gordon. 1998. The evolved psychological apparatus of human decision-making is one source of environmental problems. In T. Caro, ed., *Behavioral Ecology and Conservation Biology*, pp. 501-523. New York: Oxford University Press.

Winston, G., C. Woodbury, and G. Richard. 1991. Myopic discounting: Empirical evidence. In S. Kaish and B. Gilad, eds., *Handbook of Behavioral Economics*, Vol. 2B, pp. 325-42. Greenwich, Connecticut: JAI Press.

Winterhalder, B. 1986. Diet choice, risk, and food sharing in a stochastic environment. *Journal of Anthropological Archaeology* 5:369-392.

——. 1996a. A marginal model of tolerated theft. *Ethology and Sociobiology* 17:37-53.

——. 1996b. Social foraging and the behavioral ecology of intragroup resource transfers. *Evolutionary Anthropology* 5:2:45-57.

——. 1997. Gifts given, gifts taken: The behavioral ecology of non-market, intragroup exchange. *Journal of Archaeological Research* 5:2:121-168.

Winterhalder, B. and F. Lu. 1997. A forager-resource population ecology model and implications for indigenous conservation. *Conservation Biology* 11:6:1354-1364.

Winterhalder, B., F. Lu, and B. Tucker. 1998. Risk-sensitive adaptive tactics: Models and evidence from subsistence studies in biology and anthropology. Chapel Hill, North Carolina: University of North Carolina, Department of Anthropology. Manuscript.

Wirsing, R. 1985. The health of traditional societies and the effects of acculturation. *Current Anthropology* 26:3:303-322.

Wood, C. H. and D. Skole. 1997. Linking satellite, census and survey data to study deforestation in the Brazilian Amazon. Gainesville, Florida: Center for Latin American Studies, University of Florida. Manuscript.

World Bank. 1992. *World Development Report. Development and the Environment*. Washington, D.C.: Oxford University Press.

Yamey, B. S. 1964. The study of peasant economic systems: Some concluding comments and questions. In R. Firth and B. S. Yamey, eds., *Capital, Saving and Credit in Peasant Societies*, pp. 376-386. Chicago, Illinois: Aldine Publishing Company.

Yang, M. M. 1989. The gift economy and state power in China. *Comparative Studies in Society and History* 31:1:25-54.

Yellen, J. E. 1990. The transformation of the Kalahari Kung. *Scientific American* 262:4:96-105.

Yost, J. A. and P.M. Kelley. 1983. Shotguns, blowguns, and spears: The analysis of technological efficiency. In R. B. Hames and W. T. Vickers, eds., *Adaptive Responses of Native Amazonians*, pp. 189-224. New York: Academic Press.

Young, T. 1842. *Narrative of a Residence on the Mosquito Shore*. London: Smith, Elder.

Zimmerer, K. S. 1993. Soil erosion and labor shortage in the Andes with special reference to Bolivia, 1953-1991. Implications for 'conservation-with-development'. *World Development* 21:10:1659-1676.

Index

Printed in the USA
CPSIA information can be obtained
at www.ICGtesting.com
JSHW021437221024
72172JS00003B/35

9 780231 117852